THE HYDROGEN ECONOMY

ALSO BY JEREMY RIFKIN

Common Sense II

Own Your Own Job

Who Should Play God? (with Ted Howard)

The Emerging Order

The North Will Rise Again (with Randy Barber)

Entropy (with Ted Howard)

Algeny

Declaration of a Heretic

Time Wars

Biosphere Politics

Beyond Beef

Voting Green (with Carol Grunewald)

The End of Work

The Biotech Century

The Age of Access

The Creation of the Worldwide Energy Web and the Redistribution of Power on Earth

THE HYDROGEN ECONOMY

JEREMY RIFKIN

JEREMY P. TARCHER/PENGUIN
a member of Penguin Group (USA) Inc.
New York

Most Tarcher/Penguin books are available at special quantity discounts for bulk purchases for sales promotions, premiums, fund-raising, and educational needs. Special books or book excerpts also can be created to fit specific needs. For details, write Penguin Group (USA) Inc. Special Markets, 375 Hudson Street, New York, NY 10014.

Jeremy P. Tarcher/Penguin
a member of
Penguin Group (USA) Inc.
375 Hudson Street
New York, NY 10014
www.penguin.com

First trade paperback edition 2003

Published simultaneously in Canada

The Library of Congress catalogued the hardcover edition
as follows:

Rifkin, Jeremy.
The hydrogen economy : the creation of the worldwide
energy web and the redistribution of power on earth /
by Jeremy Rifkin.
p. cm.
Includes bibliographical references and index.
ISBN 1-58542-193-6
1. Hydrogen as fuel. 2. Globalization. I. Title.
TP359.H8R54 2002 2002025370
333.8—dc21

ISBN 1-58542-254-1 (paperback edition)

Printed in the United States of America

1 3 5 7 9 10 8 6 4 2

Book design by Lee Fukui

ACKNOWLEDGMENTS

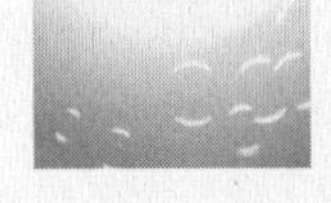

I'd like to thank my research director, Loring Katawala, for her outstanding work overseeing the project. Ms. Katawala's exceptional research skills often helped us locate difficult-to-find information, especially within the oil industry. Her attention to detail proved invaluable in dealing with the wealth of facts, figures, and statistics that make up the bulk of the book. Her enthusiasm and dedication have helped make the book a joyful experience.

I'd also like to thank C. J. Campbell, Jean Laherrère, Buzz Ivanhoe, Jim MacKenzie, John Edwards, Richard Duncan, Joel Swisher, Seth Dunn, and Brett Williams for their scientific and technical critiques of various drafts of the book as well as for their helpful suggestions, many of which were incorporated into the final pages.

Thanks also go to Ted Howard, David Helvarg, and Marty Teitel, who read early drafts of the book and made helpful suggestions.

I'd like to acknowledge my wife, Carol Grunewald, and my in-laws, Ted and Dorothy Grunewald, for many hours of fruitful conversation that helped shape my thinking during the project.

I'd like to thank Stephanie Woodhouse for her superb job in editing the various drafts of the manuscript and for overseeing the project with my publishers abroad.

I'd also like to thank Alexia Robinson for editing the book and Clara Mack for helping compile and assemble the research materials. Additional thanks go to Shreya Lamba, Kear Leng Chhour, Dara Sanandaji,

Brett Wilson, Pat Gorton, Nicole Rousseau, Tim Emmet, and Jarret Cassaniti for their help.

I'd like to thank Joel Fotinos, Cathy Fox, and Ken Siman at Tarcher/Penguin for making this project happen. I'd also like to thank my old friend Jeremy Tarcher for providing me with a unique publishing forum in which to share my ideas. His long-standing commitment to my work has made it possible for me to reach a wide public audience over the years, and it is greatly appreciated.

Finally, I'd like to extend my special thanks to my editor, Mitch Horowitz at Tarcher/Penguin, for stewarding this project along. Mitch and I spent several months going over every detail of the book. His editorial suggestions helped guide the direction of this book, and his many contributions can be found throughout the finished work.

CONTENTS

1

BETWEEN REALITIES

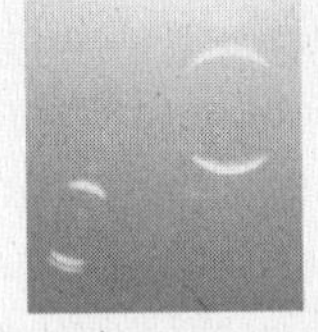

Throughout history, human beings have occasionally found themselves caught between two very different ways of perceiving reality. Certainly that was the case in the closing days of the 17th century. The Enlightenment scientists and philosophers—Isaac Newton, John Locke, René Descartes, and others—challenged many of the most revered shibboleths of church catechism, including one of its central doctrines, that the Earth is God's creation and imbued with intrinsic value. The new thinkers preferred a more materialist explanation for existence and cast their lot with mathematics and reason. Less than a century later, political renegades in the American colonies and insurrectionists in France overthrew monarchical rule in favor of a republic form of government and proclaimed man's "inalienable right to life, liberty, happiness and property." James Watt patented his steam engine on the eve of the American Revolution, consummating a relationship between coal and the new Promethean spirit of the age, and humanity made its first tentative steps into an industrial way of life that would, over the next two centuries, forever change the world.

Today, we live in similar times of great tumult, of failing orthodoxies and radical new possibilities. After two centuries of industrial production and commerce, the use of mass human labor yoked to fossil-fuel-powered machines in factories, offices, and commercial businesses is slowly falling by the wayside. New, more sophisticated and "intelligent" technologies are steadily replacing human labor in every industry and professional field. We are making a great transition to smaller, elite workforces collaborating with increasingly smart computer and robotic technologies. Within a matter of a few decades, the cheapest workers in the world will not be as cheap as the intelligent technologies that will replace them, from the factory floor to the front office. By the middle decades of the 21st century, we will likely be able to produce goods and services for everyone on Earth with only a small fraction of the human workforce we now employ. This will force us to rethink what human beings will do when they are no longer needed to labor in the marketplace.[1]

Physics and chemistry, which have dominated the era just passing, influencing every aspect of our existence, including the smallest particulars of our way of life, are making room for the age of biology. The mapping and manipulation of human, animal, and plant genomes open the door to a new era in which life itself becomes the ultimate manipulable commodity. The biotech era is beginning to raise fundamental questions about the nature of human nature, and the public is quickly being swept up in a great debate between those who view the new age as a biological renaissance and others who warn of the coming of a commercial eugenics civilization.[2]

The computer and the telecommunications revolution have given birth to the Internet and the World Wide Web, marking a great change in the way human beings communicate. "Access" has become the all-encompassing metaphor for a generation of people who can now connect with one another via an electronically mediated "central nervous system" that spans the globe. The new "speed of light" society is changing the way we conduct business. The market economy, steeped in the exchange of goods and services between sellers and buyers, is found to be far too slow to accommodate the new speed of commercial life. In the coming era, the exchange of property in markets steadily gives way to access to services and experiences in networks. In a society where time itself is the most scarce and valuable resource, suppliers retain ownership of property and users pay for the time they spend accessing goods and services. Subscriptions, leases, time-shares, licenses, and rentals become the preferred way of doing business. The new "temporal" economy is characterized by falling transaction costs and diminishing profit margins, forcing commercial enterprises to introduce a radical new business model based on "shared savings" arrangements among network partners. The transformations from property exchanges to access relationships and from profit margins to shared savings are beginning to restructure commercial life around the world.

Our notions of what constitutes culture is also radically changing. Giant content companies like Disney, Universal Vivendi, AOL–Time Warner, and Sony are mining cultural resources all over the world, transforming them into paid-for experiences of every kind. The high-end income earners—the top 20 percent of the world's consumers—now spend almost as much money on experiences as on basic goods and services.

A younger generation of cultural activists who oppose the new commerce are waging an escalating battle against "branding," lifestyle marketing, and new kinds of retail franchising and entertainments, all of which they believe are leading to the homogenization of culture. They argue that the new global cultural commerce is a threat to the world's cultural diversity, and they seek protection of indigenous cultures. The commercial sphere's effort to subsume the cultural sphere and become the sole arbiter of the human story represents a great turning point in the relationship between commerce and culture with profound long-term consequences for every society.[3]

The transformation in the nature of work, the emerging biotech and communications revolutions, the growing temporalization of economic activity, and the global struggle between commerce and culture are fundamentally altering both the conception and reality of the world around us.

Now, an equally profound change is about to occur in the way we use energy. The modern age was made possible by the harnessing of coal, oil, and natural gas. All of the advances of the past two centuries, whether they be commercial, political, or social in nature, are connected, in some way, to the massive power surge unleashed by the burning of fossil fuels.

Anthropologists say that the amount of energy consumed per capita in a society is a good measure of its relative state of advance. Western society, over the last 200 years, has consumed more energy per capita than all other societies throughout all of recorded history put together. We have come to enjoy an unprecedented standard of living, and we owe our good fortune to fossil-fuel deposits formed millions of years ago. Manna, yes! But, not from heaven, but rather from deep beneath the Earth.

Alas, all good fortunes eventually come to an end. For a long time, we naively believed that, if the supply wasn't unlimited, there was at least enough oil stored away in the nooks and crannies of the Earth to supply all of our needs far into the foreseeable future. When oil production in the United States peaked in 1970—the point at which half of all the recoverable reserves had been produced—geologists became uneasy. But, as long as oil continued to flow from other parts of the world, the average American did not give the matter a passing thought. It wasn't until the Arab oil embargo, three years later, that Americans and consumers in other countries took notice. Waiting in line at filling stations for hours at a time in hope of securing a few gallons of gasoline was a sobering experience for millions of people. At the time, some critics warned that we would soon run out of oil. But it didn't happen. The United States, and other nations, as well as leading energy companies, began a frantic search for new sources of oil, and they found them. The lines at the gas pump dwindled and the oil crisis abated. The gasoline flowed and it was cheaper than ever. The world got back to business as usual.

At present, oil is relatively cheap on the world markets. Our experts tell us that while oil—and natural gas—will eventually begin to run out, that time is at least thirty to forty years away, maybe even longer—plenty of time to plan for alternative sources of energy.

But what if you were suddenly told that everything is not exactly as it seems on the oil front? Imagine waking up very soon to a headline in the newspaper: GLOBAL OIL PRODUCTION PEAKS; PRICES EXPECTED TO RISE DRAMATICALLY ON WORLD MARKETS IN COMING YEARS.

A growing number of the world's leading geologists are beginning to say just that. This time, they warn, there really is an oil crisis looming on the horizon, and when it hits it will be permanent. So where are we headed in the next few years?

If global oil production were to peak some time in the next decade or so, followed shortly thereafter by a global peak in natural-gas production, it could set off a cascade of events that could unravel much of our industrial way of life. Two developments, in particular, are likely to figure prominently in the coming oil equation.

First, while the experts disagree about when global oil production is likely to peak, they agree that when it does, virtually all of the remaining untapped reserves will be left in the Muslim countries of the Middle East, potentially changing the current power balance in the world. The juxtaposition of dwindling oil reserves and growing militancy among many of the world's younger Muslim population could threaten the economic and political stability of every nation on Earth. Political leaders and policy analysts are particularly worried about the escalating conflict between the Israelis and the Palestinians and the possibility that, in the future, Islamic fundamentalists might pressure their governments to use oil as a weapon against the United States and other Western nations for supporting the Israelis.

Second, if global oil and natural-gas production peak, catching the world unprepared, countries and energy companies are likely to look to the dirtier fossil fuels—coal, heavy oil, and tar sand—as substitutes. With Earth's temperature already projected to rise by 2.52 to 10.44°F between now and the 22nd century, the switch to dirtier fuels would mean an increase in CO_2 emissions, a greater temperature rise than is now being forecast, and even more devastating effects on the Earth's biosphere than have already been envisioned.[4]

Ours is not the first great civilization in history to face an energy crisis. Energy has played an important role in the rise and fall of civilizations. Many anthropologists and historians would argue that it has most often been the critical factor. If there are lessons to be learned—and there are—about how other civilizations have come to grips with their own energy crises, now would be the time to take stock, argue the Cassandras. The fact is, there are ironclad rules that govern the flow of energy, and if those rules are breached, civilizations can perish. The laws of thermodynamics tell us what, in the final analysis, the upper limits are in the pursuit of human power over the environment. Societies that reach beyond the constraints imposed by their own energy regimes risk breakdown.

The United States, and every other country, is vulnerable to mounting internal and external threats and disruptions as we move into the last stages of the oil era. Our vulnerability is even more pronounced because of the highly centralized, hierarchical energy infrastructure, and the accompanying economic infrastructure, that we created to manage a fossil-fuel energy regime. The fossil-fuel era is characterized by a top-down organizational scheme made necessary by the difficulty of harnessing and exploiting hard-to-find forms of energy. The extraordinary costs associated with the processing of coal, oil, and natural gas required vast amounts of investment capital and led to the formation of giant energy enterprises. Currently, ten to twelve mega-companies, both commercial and state-owned, dictate the terms by which energy flows through the whole world. By centralizing power over the Earth's energy resources, the energy companies created the conditions that rewarded economies of scale, and centralization of economic activity, in every other industry.

The burning of fossil fuels also quickened commercial life. Managing the increased density and flow of human commerce further solidified the formation of highly centralized, hierarchical commercial enterprises. Today, fewer than 500 global companies control most of the economic activity of the planet. Globalization represents the end stage of the fossil-fuel era—a period in which fewer and fewer corporate institutions micro-manage both the flow of energy and economic activity in communities throughout the world.

Globalization is the defining dynamic of our time. Proponents look to it as the next great economic advance for humanity and as a way to improve the lives of people everywhere. Its critics view it as the ultimate example of corporate dominance over the affairs of society and as a means to deepen the gap between the haves and have-nots. Transnational corporations, with the help of the G–7 nations, are lobbying to change government regulations and statutes that, they argue, restrict freedom of trade. Anti-globalists are taking to the streets in greater numbers to protest what they contend is the systematic gutting of environmental and labor standards designed to protect the Earth's ecological and human communities from corporate rapacity. The tragic events of September 11th, and its aftermath, have exacerbated the tensions surrounding globalization and have made everyone feel more vulnerable in what is becoming an increasingly unsafe and unsure world.

Despite the growing dissension and polarization, little serious effort has been made to analyze the critical underlying factors that have led to globalization and the acrimonious reactions to it. While globalization can be understood from a number of different perspectives, none is more important than the energy equation. We sometimes forget that without fossil fuels globalization would have been impossible. Fossil-fuel energy has allowed commercial enterprises to radically compress time and shorten distances, making possible a single world market for the exploitation of raw resources and human labor and for the marketing of finished products and services.

It is no wonder that controlling fossil-fuel energy reserves has been the central preoccupation of governments and industry for more than a century. Geopolitics has been, to a great extent, synonymous with the politics of oil for five generations. Those countries, companies, and peoples that have successfully controlled the flow of oil have enjoyed unparalleled wealth, while those who have been denied favored access to the wealth-generating potential of what geologists call "black gold" have slipped further into poverty and have been the subject of increasing exploitation and marginalization.

Consider, for example, the rise in oil prices in the 1970s and '80s, which was a major cause of the escalating debt crisis in Third World countries. Unable to afford the high price of oil on world markets, developing nations were forced to secure billions of dollars in commercial and institutional loans to pay for more expensive oil imports and the increasing costs of all the other activities associated with higher oil bills. The debt burden has worsened in recent years as developing countries have become even more dependent on foreign oil to modernize their manufacturing economies and meet the needs of growing urban populations. Many of the world's poorest countries are now spending more money to pay back past loans than they are spending on basic human services. The result is an irreversible downward cycle of ever-deepening poverty and despair. Protesters at recent global-development forums have zeroed in on the Third World debt crisis as the most visible sign of the inequities created by globalization and are demanding a cancellation of debts owed by poor countries. The fossil-fuel energy regime, then, is both the vital force that makes globalization possible and one of the factors most responsible for the growing divide between the rich and poor nations of the world.

Now, however, the global infrastructure created to exploit fossil fuels and manage industrial activity is aging and beginning to crack at the seams. The fissures are everywhere, and there is mounting concern that the infrastructure might not hold together much longer. Some geologists are beginning to suggest that the system itself could collapse. To be unprepared for what might come, say the worriers, would be foolhardy.

But what does "prepared" really mean? If the fossil-fuel era is passing, what can replace it? A new energy regime lies before us whose nature and character are as different from that of fossil fuels as the latter was different from the wood-burning energy that proceeded it.

Hydrogen is the lightest and most ubiquitous element found in the universe. When harnessed as a form of energy, it becomes "the forever fuel."[5] It never runs out, and, because it contains not a single carbon atom, it emits no carbon dioxide. Hydrogen is found everywhere on Earth, in water, fossil fuels, and all living things. Yet, it rarely exists free-floating in nature. Instead, it has to be extracted from natural sources.

The foundation is already being laid for the hydrogen economy. In the next few years, the computer-and-telecommunications revolution is going to fuse with the new hydrogen-energy revolution, making for a powerful mix that could fundamentally reconfigure human relationships in the 21st and 22nd centuries. Since hydrogen is found everywhere and is inexhaustible if properly harnessed, every human being on Earth could be "empowered," making hydrogen energy the first truly democratic energy regime in history.

Hydrogen-powered fuel cells are being commercially produced for installation in factories, offices, commercial buildings, and homes to produce power, light, and heat. The major automakers have spent more than $2 billion developing hydrogen-fuel-cell-powered cars, buses, and trucks, and they expect to mass-produce the new generation of vehicles before the end of the current decade. Locating micro-power plants on-site with the end user—so-called "distributed generation"—threatens the long-standing dominion enjoyed by centralized power plants that grew up with the fossil-fuel era. Now, the end user becomes his or her own producer as well as consumer of energy. When millions of small power plants are connected into vast energy webs, using the same architectural design principles and smart technologies that made possible the World Wide Web, people can share energy and sell it to one another—peer-to-peer energy sharing—and break the hold of giant energy and power companies forever.

The worldwide hydrogen energy web (HEW) will be the next great technological, commercial, and social revolution in history. It will follow on the heels of the development of the worldwide communications web in the 1990s and, like the former, will bring with it a new culture of engagement. While the HEW is potentially a revolution in energy design that could decentralize and democratize energy and recast commercial and social institutions along radically new lines, there is no guarantee that, in fact, it will. Here, the history of the Internet and the World Wide Web is instructional. The Internet contains the promise of empowering billions of people, giving everyone on Earth potential access to everyone else and making communication and exchange of information between people truly democratic. The "Net" activists of the 1990s argued that information ought to be freely shared. While community nets and free-nets were established early to make good on that vision, they were too few, too weak, and too devoid of meaningful content to withstand a better-financed, more highly organized campaign waged by companies like AOL and

Microsoft to gain control over the new medium. Commercial forces have conspired, from the very beginning, to gain an unbreakable hold over the portals of cyberspace so that they could become the gatekeepers and arbiters of the Information Age. A similar threat and challenge face the hydrogen energy web.

Whether hydrogen becomes "the people's energy" depends, to a large extent, on how it is harnessed in the early stages of development. Like the Net activists of the last decade, a new generation of energy activists is beginning to argue that hydrogen energy ought to be shared. Making that happen will require that public institutions and nonprofit organizations—especially the publicly owned utilities that provide energy to hundreds of millions of people and the thousands of nonprofit cooperatives with a collective worldwide membership that exceeds 750 million people—jump in at the beginning of the new energy revolution and help establish distributed-generation associations (DGAs) in every country.

Connecting the whole of humanity into hydrogen energy webs is going to require the active participation of the private sector as well. Commercial enterprises will develop and manufacture the new hardware and software of the distributed-generation revolution and will play a big role in bundling energy services and coordinating the flow of energy on the hydrogen energy web. Creating the appropriate partnership between commercial and non-commercial interests will be critical to establishing the legitimacy, effectiveness, and long-term viability of the new energy regime.

The switch to a hydrogen economy can end the world's reliance on imported oil and can help diffuse the dangerous geopolitical game being played out between Muslim militants (in the Middle East and elsewhere) and Western powers. Equally important, weaning the world away from a fossil-fuel energy regime will limit CO_2 emissions to only twice their preindustrial levels and mitigate the effects of global warming on the Earth's already beleaguered biosphere.

A decentralized, hydrogen-energy regime offers the hope, at least, of connecting the unconnected and empowering the powerless. When that happens, we could entertain the very real possibility of "reglobalization," this time from the bottom up, and with everyone participating in the process.

The fossil-fuel era brought with it new ways of organizing society, including industrial enterprise, nation-state governance, dense urban settlement, and a bourgeois lifestyle. Because it is so different from the various forms of hydrocarbon energy, hydrogen will give rise to a wholly new type of energy infrastructure as well as to radically different economic institutions and new patterns of human settlement, just as coal and the steam engine and, later, oil and the internal combustion engine did in the past. When every human being on Earth can be the producer of his or her own energy, the very nature of commercial life radically changes. Economic activity becomes far more widely diffused. The disaggregation of commerce, in turn, allows for a disaggregation of human settlement. The centralization of power and economies of scale that so characterized the fossil-fuel era inevitably led to a concentration of human population in mega-cities that used up vast amounts of energy and were ultimately unsustainable. The creation of decentralized hydrogen energy webs connecting end users would make possible the establishment of human settlements that are more widely dispersed and more sustainable in relationship to local and regional environmental resources.

The worldwide hydrogen energy web, like the worldwide communications web, will allow us to connect every human being on the planet with every other in an indivisible and interdependent economic and social matrix. The human species can now become a human community fully integrated into the Earth's ecosystems. Unfortunately, our ideas about personal and collective security are still mired in a fossil-fuel state of mind. In the oil age, each human being's sense of personal security came to mirror the organizational values of the larger institutional framework that managed the flow of energy and economic activity. Autonomy and mobility became the undisputed social virtues of the era, in both personal and institutional life. In the coming hydrogen economy, the sheer density of human interaction, as well as the speed of engagement, will give rise to a new sense of security, bound up in embeddedness in multiple commercial, social, and environmental networks and in global interdependence. Our individual security and the well-being of the Earth's diverse human, biological, and geological communities will become seamless. We will come to see ourselves as part of a single Earth organism.

The divisive geopolitics that so permeated the fossil-fuel era will give way to a new sense of biosphere politics in a hydrogen age.

We find ourselves on the cusp of a new epoch in history, where every possibility is still an option. Hydrogen, the very stuff of the stars and our own sun, is now being seized by human ingenuity and harnessed for human ends. Charting the right course at the very beginning of the journey is essential if we are to make the great promise of a hydrogen age a viable reality for our children and a worthy legacy for the generations that will come after us.

2

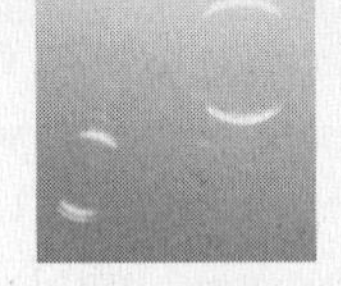

SLIDING DOWN HUBBERT'S BELL CURVE

The energy crisis of the 1970s is a fading memory. We appear to be awash in oil. Oil prices in the spring of 2002 hovered at $24 per barrel on world markets, and OPEC countries struggled to maintain market share with non-OPEC producers—including Russia, Mexico, and Norway—who produce an increasing amount of the world's petroleum.[1] Americans are on the road with millions of gas-guzzling cars, prices at the pump are rising but manageable, and the cost of heating oil is flat. Talk of energy conservation, once a national pastime, is now heard only infrequently, both here in the United States and in other industrialized nations.

Our policy leaders tell us that, thanks to new exploratory technologies, we are finding more oil to replenish the stores and that, because of new drilling technologies, we are securing more hard-to-extract oil from existing fields. According to the Energy Information Administration (EIA) of the U.S. Department of Energy, global peak production for cheap crude oil is nearly thirty-five years away, plenty of time to make the transition to alternative energy strategies. In short, while the world faces many problems in the years ahead, a shortage of oil is not one of them. The energy foundation for the industrial and post-industrial way of life, we are told, is secure.

Amid the apparent complacency, however, a number of the world's leading geologists and oil consultants are publishing the results of new studies that tell a much different story. Their calculations suggest that global production of cheap crude oil—the lifeblood of the global economy—could peak before the year 2010, but no later than 2020.[2] ("Peak" is believed to occur when about half of the estimated ultimately recoverable reserves [EUR] of oil in the world have been produced.) While these new highly controversial studies have been published in the world's leading scientific journals, including *Science* and *Scientific American*, and have sparked a lively debate within the field of petroleum geology and in the corporate boardrooms of some of the world's leading energy companies, they have yet to be reported widely in the media. Our politicians and policy makers are largely unaware of the new data, and our economists and business leaders remain equally uninformed. Yet, if these studies prove to be accurate, we are fast approaching one of the great historical crossroads for human civilization, with far-reaching impacts whose extent we can only begin to grasp.

Cooking the Books

When geologists talk about the global peak in oil production, they are referring primarily to what is known as "conventional oil" or "light oil," the kind that gushes freely from beneath the ground on land or offshore and is easily transformed into gasoline and other petroleum-based products. There are also "unconventional oils" (oil from tar sand, heavy oil, oil from deep water or polar regions, and oil shale).[3]

Oil is composed of organic material, most of which is derived from floating planktonic plants, including green algae, and from single-celled planktonic animals.[4] The organic debris settled to the bottom of lake beds and oceans, where it was preserved from oxidation by stagnant water. After burial, the plankton was converted by heat and pressure into oil and gases, later to be used as source beds. Much of the oil deposits were formed in the late Jurassic period, more than 150 million years ago, in the tropical regions near the equator. The source rocks then migrated, with shifts in the plate-tectonic movements of the continents, north and east to the Middle East, the North Sea, Siberia, and other northern areas. The oil deposits in the United States date back to the Permian era (230 million years ago), and the oil in Venezuela was formed in the Cretaceous era (90 million years ago).[5]

Geologists agree that more than 875 billion barrels of oil (Gbo) have been removed from the Earth thus far, almost all of it in the past 140 years of the industrial era.[6] What they don't agree on is how much conventional oil is still left to be produced. (It should be pointed out that even though the experts quibble over the amount of oil left, their figures nonetheless lie within a fairly tight range). Some of the problem is attributable to the various interpretations used to define "reserves."

First, geologists and engineers distinguish between reserves and resources. "Reserves" refers to the known quantity of oil that lies in fields and that can be produced with existing technologies, within a foreseeable time frame, at a commercially reasonable cost. "Resources" refers to theoretical estimates of the total amount of oil that may exist in a region, including that which can't be economically extracted or processed with existing technologies or under current market conditions.[7] Complicating the matter, the industry also relies on a number of other terms to define "reserves," including "active" and "inactive," "probable," "possible," "inferred," "identified," and "undiscovered."[8]

Veteran geologist Jean H. Laherrère says that this proliferation of qualifying terms for designating reserves is intentional and designed to allow countries and companies to fudge the figures—a kind of creative geological accounting—for political and commercial purposes.

Laherrère blames "vested interests which make use of lax definitions to propose numbers that meet their political objectives: oil is money and reserves are, so to speak, oil in the bank—a bank far underground where no auditors can check the account."[9]

Oil shale is a good example of the "creative accounting" practices. The U.S. government often cites the fact that there are enough oil shale "resources" to produce two trillion barrels of oil, giving the impression that the U.S. has an ample supply of fossil-fuel "reserves" to maintain its energy independence. Yet, no shale oil has been commercially produced, because it is not economically recoverable with existing mining and refining techniques. So, while shale is a resource, it does not qualify as a reserve.

For the purposes of this discussion, three figures are critical: the cumulative production (how much conventional oil has been produced globally to date); an estimate of global oil reserves; and a projection of recoverable oil yet to be discovered. These three factors added together constitute the ultimate total of recoverable oil.

As already mentioned, oil exists in basins in which organic materials were deposited and preserved. These basins are on land and in shallow offshore waters. Geologists have located 600 such basins, and it is widely believed that there are few left to discover.[10] Four hundred of these basins have already been explored. The other 200 are located in remote areas such as Greenland and in deep ocean waters off Brazil, West Africa, and the Gulf of Mexico, and are both difficult and costly to explore. Significant amounts of oil have been found in 125 basins.[11]

Geologists traditionally determined the amount of undiscovered conventional oil reserves that exists by the following method: First, they calculated how much oil had been secured in the 400 basins where oil had already been drilled and ascertained how many cubic miles of sediment that oil had been found in. By dividing one figure into the other, geologists calculated a world average of how much oil might be expected to exist per cubic mile of sediment. Then they calculated the volume of the sediment in all of the basins in the world and multiplied it by the average amount of oil found per cubic unit. Today, using highly sophisticated geochemical modeling, geologists can make far more accurate estimates of how much oil is left to access.

The United States (the lower 48 states), which is estimated to have approximately 195 billion barrels of ultimately recoverable reserves, has already produced 169 billion barrels of its largesse, leaving it with reserves of only 20 billion barrels and yet-to-find reserves of approximately 6 billion barrels.[12] Saudi Arabia, in comparison, has ultimately recoverable reserves of 300 billion barrels. The Saudis have only produced 91 billion barrels, leaving its reserves at 194 billion barrels and yet-to-find reserves of an additional 14 billion barrels. Russia's ultimately recoverable reserves are approximately 200 billion barrels. Russia has already produced 121 billion barrels of its total, leaving it with reserves of only 66 billion barrels and yet-to-find reserves of 13 billion barrels. So while the U.S. has only 14 percent of its original oil still unproduced and Russia has 39 percent of its oil still untapped, Saudi Arabia has 70 percent of its oil still in the ground.[13]

Unfortunately, according to a growing chorus of skeptics, the oil figures in the official literature are suspect, not because the science isn't available to get an accurate read but rather because, as Laherrère suggests, countries are cooking the books to inflate the figures.[14]

According to the U.S. Geological Survey, there are 3,003 billion barrels of estimated ultimately recoverable reserves (EUR). Some of the new computer models, however, suggest that the EUR figures are only two-thirds as much as those claimed by the USGS.[15] The different estimates, while not very significant in the long sweep of human history and even less so in terms of geological history, become quite important in the geopolitical arena, where the impact of change is measured in years and decades. Using the USGS figure of 3,003 billion barrels of ultimately recoverable reserves and assuming the current annual production growth rate of 2 percent, the EIA estimates that the global production of oil will peak in 2037. But, if the new models are correct, the time line to global peak oil production narrows to between eight and eighteen years. (Even the EIA's own analysis, based on current USGS figures on ultimately recoverable oil and using slightly different assumptions in the modeling process, acknowledges that the peak in global production could come as early as 2016.)[16] While the varying opinions reflect legitimate differences in interpretation of data, there is also the question as to what extent official figures have been subject to political and commercial pressures. Let's look at the evidence.

Reserve figures are published yearly by the *Oil and Gas Journal* and *World Oil*. These trade journals survey each country and list its statistics without any means of independent verification. The result is a hodgepodge of numbers that are often misleading.

For example, geologists assign probabilities to their assessments of recoverable reserves. Geologists Colin Campbell and Jean Laherrère cite the case of the Oseberg field in Norway. Petroleum engineers estimate that there is a 90 percent chance that this particular field will yield 700 million barrels of recoverable oil, but only a 10 percent chance that it will yield 2,500 million more barrels. The lower figure is called the P^{90} estimate and the higher figure the P^{10} estimate.[17] In the United States, the Securities and Exchange Commission only allows companies to refer to reserves as proved "if the oil lies near a producing well and there is 'reasonable certainty' that it can be recovered profitably at current oil prices, using existing technology."[18] This is a P^{90} estimate. Campbell and Laherrère argue that a P^{90} estimate is too narrow and often underestimates the amount of oil that will be produced during the life of a field. They believe that a medium estimate, called a "proved and probable," or P^{50}, is more accurate. The P^{50} estimate "is the number of barrels of oil that are as likely as not to come out of a well during its lifetime, assuming prices remain within a limited range."[19]

If the U.S. underestimates proved reserves by using a P^{90} estimate, countries like the former Soviet Union have consistently overestimated proved reserves by using P^{10} assumptions. The Russian government routinely overstates reserves by including geological reserves—everything that exists in situ—as opposed to the concept of "economically producible reserves."[20] The gap between fact and fiction can become so large as to be absurd. In 1996, *World Oil* listed the former Soviet Union as having proved reserves of 190 Gbo. That same year, the *Oil and Gas Journal* listed proved reserves in the former Soviet Union as 57 Gbo.[21]

OPEC countries inflate the figures, say the critics, in order to increase their production quotas and to secure international loans from lending institutions like the World Bank and the International Monetary Fund (IMF) or to leverage private bank loans for infrastructure development and new commercial ventures.

To get a sense of exactly how fast and loose the reporting can be, consider the fact that in the mid-1980s the reported global proved crude reserves were listed between 650 billion and 700 billion barrels. By the 1990s, 300 billion additional barrels of crude oil had been added to global proved reserves—one-third of all world reserves—despite the fact that no significant new oil fields had been discovered. Virtually all of the increase came from OPEC countries. Saudi Arabia, whose proved reserves hovered between 163 billion and 170 billion barrels for years, suddenly jumped to 257.5 billion barrels in the year 1990 alone. Kuwait's reserves jumped 26 billion barrels—from 63.9 to 90 billion—between 1984 and 1985. Iraq's proved reserves more than doubled between 1987 and 1988, from 47.1 billion to 100 billion. Iran went from 48.8 billion in 1987 to 92.9 billion twelve months later.[22] Abu Dhabi and Dubai reported a tripling of their reserves in 1988.[23] In 1988, three major oil-producing countries claimed twice the reserves they had just a year before. It is true that the earlier numbers were probably too low and needed to be adjusted upward. That is because when foreign companies owned the oil fields they often underreported new discoveries to reduce their taxes. Still, the monumental increases in reported reserves in the mid- to late 1980s far exceeded what could be justified by updating earlier underreported corporate figures.

Adding to the chicanery, while oil companies discovered an average of seven Gbo a year in the 1990s, they drained more than three times as much oil.[24] Still, year after year throughout the 1990s, more than half the countries listed in the *Oil and Gas Journal* yearly report claimed to have the same proved reserves as they did the year before.[25] In 1997, fifty-nine oil-producing nations reported that their reserves were unchanged from the year before, even though existing fields were being pumped for oil and at least some new fields were being discovered. By 1999, the number of countries reporting unchanged reserve figures had increased to seventy.[26]

Reality Check

So, how much cheap crude oil is still out there and available to be recovered? With the introduction of 3-D digital seismic methods in the late 1960s, geologists have been able to pinpoint new oil fields with increasing accuracy. Global exploration for oil greatly intensified in the 1970s and 1980s in the aftermath of the Arab-Israeli war and the subsequent OPEC oil embargo, and later the Iraq-Iran war, which sent oil prices soaring to $40 per barrel.[27] The U.S. and other countries, as well as global energy companies, were worried about their increasing dependency on Mid-East oil and determined to find alternative sources. Oil companies began a worldwide search for new fields. In the U.S. alone, the number of exploratory and development wells increased from 28,000 to 90,000 between 1973 and 1981.[28] Yet, despite the mammoth effort, proved reserves in the lower forty-eight states actually declined during that period, from 25 billion barrels in 1973 to 20 billion barrels in 1986, and crude oil production declined by 24 percent.[29]

Similar disappointing results were reported around the world. Geologists now concede that most of the major oil fields had already been found before the newest wave of exploration. The economists' long-standing argument that rising oil prices would have the effect of stimulating new exploration and lead to significant new oil finds was proved questionable.

There are 1,500 major and giant oil fields in the world today. They contain 94 percent of all the known crude oil. The 400 largest fields contain 60 to 70 percent of the total. Only forty-one of these fields have been found since 1980.[30] The bottom line, says Colin Campbell of the Oil Depletion Analysis Center of London (ODAC), is that, "by now, the whole world has been thoroughly explored, so it has become clear that no new provinces comparable with the North Sea and Alaska await discovery."[31] The U.S. Geological Survey concurs. It reports that global discovery of new oil fields peaked in 1962 and has been declining ever since. The golden age of oil has already passed. That is not to say that new smaller fields won't continue to be found, but they will not be able to offset the continued decline in the world's proved reserve inventory.[32]

All of this is disturbing news when we consider that world demand for crude oil is now about 24 billion barrels per year and rising but that we're only finding less than twelve billion barrels of recoverable oil in new fields each year, and even that number is expected to decline with each passing year.[33] In other words, we are consuming two barrels of conventional crude oil for every new barrel discovered.[34]

Although heads of states, policymakers, and economists in the West continue to talk enthusiastically about increasing oil production in existing fields outside the Middle East, the reality is that the boost in non-OPEC oil production over the past twenty years is already showing signs of petering out. Much of the new oil in the 1980s and 1990s came from the North Sea. The Energy Intelligence Group estimates that North Sea production will peak by the end of 2002 at 6.77 million barrels.[35] The Caspian Sea is another area where geologists and energy companies had hoped to find an oil bonanza on the scale of a mini-Persian Gulf. In reality, the area contains approximately 50 Gbo of ultimately recoverable reserves, about the same as the North Sea, and is expected to peak in production by 2010.[36] Meanwhile, Russian oil companies are feverishly adding new refineries and pipelines and dumping more oil onto world markets, and this is having the effects of lowering world oil prices and depleting Russia's proven reserves.[37]

Like the North Sea and the Caspian Sea, the Alaskan oil fields are the subject of much speculation among oil men, resource economists, and political leaders. The Bush Administration has been keen on opening up the Arctic National Wildlife Refuge (ANWR)—a pristine wildlife habitat protected by law from commercial exploitation—for oil exploration. The additional oil that might flow from the region, however, is so little as to be largely irrelevant in addressing the question of U.S. energy independence.

While the U.S. Geological Survey estimates that 20.7 billion barrels of recoverable oil may lie beneath the ANWR, the amount of recoverable oil using existing technologies is significantly less—7.7 billion barrels at most. The amount of commercially producible oil—the amount recoverable at $20 per barrel—is even less, about 3 billion barrels. This means that the technically recoverable amount is only equivalent to 390 days of conventional oil at the current rates of use, and that the economically recoverable amount is equivalent to only 152 days of oil use. The Energy Information Administration (EIA) of the U.S. Department of Energy forecasts that by 2020, ANWR could provide 1.4 million barrels per day. But, with oil production in 2020 expected to hover around 112 to 120 million barrels per day, ANWR would be adding only about 1 percent to the global oil supply.[38]

Franco Bernabe, formerly senior economist at the Paris-based Organization for Economic Cooperation and Development and CEO of the Italian oil company ENISpA, recently surveyed 200 leading non-OPEC oil companies and found that: "From 1980 to 1997, their reserve-to-production ratio declined from eighteen years to twelve years."[39] This is bad news for the future of non-OPEC oil production. Bernabe concludes that "even to maintain this ratio at today's 2.5 percent annual increase in world production, this group would need to replace 140 percent of their reserves over the next five years," a virtually impossible task.[40]

The decline in new discoveries and the depletion of proved reserves becomes even more significant in light of the projected increase in world demand for oil over the next two decades. With the world population expected to increase from 6.2 billion to 7.5 billion by the year 2020, the pressure on oil reserves is only going to intensify.[41] Increasing population will bring with it a speeding up of the process of urbanization. That means more oil for transportation, heating, electricity, and agricultural and industrial production. The energy needs of a burgeoning population are going to exert unprecedented strain on the world's remaining crude-oil reserves.

It's illusory even to suggest that the exploding populations of the developing nations will ever be able to have access to the amount of oil per capita that the U.S. has enjoyed during the golden age of oil. If China were to use as much oil per capita as we do in the U.S. to maintain our standard of living, it would need to consume 81 million barrels of oil a day—that's 10 million more barrels than the entire world oil production in 1997.[42] Even if countries like China and India were to increase their energy consumption to the per-capita level of South Korea, says *Fortune* magazine, "these two countries alone will need a total of 119 million barrels of oil a day. That's almost 50 percent more than the world's entire demand in the year 2000."[43]

China and India are bellwethers of a great transformation taking place in the developing world. Poorer countries are industrializing, urbanizing, and modernizing in a race to catch up with the standard of living enjoyed by the industrialized countries of the West. They need oil, and because oil is becoming more scarce, they are going to find themselves locked in fierce competition with the industrialized nations of the North for access to the remaining reserves. Writing in *The Economist*, former business editor Edward Carr estimates that

> by 2010 the share of total energy consumption accounted for by the rich countries will have fallen below 50% for the first time in the industrial era. . . . The growth in energy consumption in developing countries between 2000 and 2010 will be greater than today's consumption in Western Europe.[44]

The rising demand for oil, in both the industrialized and the developing world, is likely going to be the critical factor in the geopolitical struggles of the first quarter of the 21st century. The EIA projections on global oil demand reveal how arduous the challenges are that lie ahead. According to the agency, world daily oil demand will increase from 80 million to 120 million barrels per day by 2020, a 50 percent increase in less than twenty years.[45] Finding and producing 40 million additional barrels of cheap oil per day is going to prove difficult.

It should be emphasized that the issue is not whether we are running out of oil, but whether we are nearing peak production of oil, which has been the lubricant for the industrial advances of the 20th century. On this score, there are differences of opinion among the experts.

Interestingly, most geological surveys conducted over the past fifty years have been remarkably consistent in their estimates of ultimately recoverable global oil. James J. MacKenzie, former senior staff member for energy at the President's Council on Environmental

Quality (CEQ) from 1977 to 1981 and currently senior associate at the World Resources Institute's Program on Climate, Energy and Pollution, writes that "the great majority of these studies reflect a consensus among oil experts that the EUR for oil lies within the range of 1,800 to 2,200 billion barrels."[46] The world has already consumed more than 875 billion barrels of the total.

As mentioned earlier, the newest U.S. Geological Survey puts the EUR for oil higher than earlier forecasts, at 3,000 billion barrels. Its optimistic projection is based, in part, on the belief that the former Soviet Union, the Middle East, the Niger and Congo deltas of Africa, and the northeast Greenland shelf contain potentially large, unexplored reserves.[47]

John Edwards of the University of Colorado is equally optimistic. His projections are based on adding unconventional oil, including Venezuelan heavy oil and Canadian tar sands, as well as the possible conversion of 20 percent of the natural-gas reserves to liquids to the EUR estimates. Using a more liberal interpretation of what constitutes the EUR, Edwards predicts a global peak in oil production between the years 2030 and 2040.[48] If just conventional oil is being considered, Edwards forecasts a peak in global production between the years 2020 and 2030.[49]

Colin Campbell and Jean Laherrère, however, come in at the lower end of the range, putting the total EUR for conventional crude oil at only 1,800 billion barrels.[50] They argue that oil-producing countries, especially OPEC countries and Russia, have grossly inflated their reserve figures for political ends. Campbell and Laherrère are not alone. A growing number of the world's leading geologists are publishing the results of new computer-modeled studies, and their findings differ substantially from the conventional wisdom.

These new studies suggest that global oil production will peak sometime between 2010 and 2020—and a few of the studies estimate a peak even earlier than 2010. In other words, within that time frame, half of the ultimately recoverable reserves of oil will have been produced. Once global oil production peaks, oil prices will begin a steady, uninterrupted rise as nations, businesses, and consumers vie for the remaining half of the reserves. Unlike the first oil shock in the 1970s and 1980s, which was politically induced, this time the shock will be based on real shortages. With each passing year, there will be less cheap crude oil available in the world. Diminishing cheap crude oil and growing human population (especially in the developing world) will create a dangerous new dynamic.

The modeling methodology being relied on to make these predictions is called the Hubbert Bell Curve. M. King Hubbert was a geophysicist who worked for the Shell Oil Company. In 1956, he published what has since become a famous paper predicting the peak and decline of oil production in the lower forty-eight states. Based on the amount and rate of past production, he projected that oil production in the U.S. would peak between 1965 and 1970. At the time that he made this forecast, America was producing a record amount of oil. Most geologists and energy company executives scoffed at the prediction, ridiculing the author and his thesis. To their surprise, Hubbert's prediction came true. Production peaked in 1970 and began its steady decline. The United States lost its preeminence as the world's leading producer of oil, and that change has dictated much of the geopolitics of the world ever since.

Hubbert's thesis is elegant in its simplicity. He argued that oil production starts at zero, rises, peaks when half the estimated ultimately recoverable oil is produced, and then falls, all along a classic bell-shaped curve. The extraction of oil begins slowly and then begins to accelerate quickly as big oil fields are located. After the biggest fields are found and exploited, production begins to slow. The smaller fields become harder to find and the oil more expensive to drill and produce. At the same time, as the larger fields are drained, the remaining oil becomes more difficult to pump to the surface. The gush gives way to a slow, steady, declining flow. The combination of the declining rate of discovery and the decline in production from existing fields eventually results in a production peak. The top of the bell curve represents the midpoint, at which half of the ultimately recoverable reserves have been produced. From that point on, production drops as fast as it has climbed, along the second half of the bell curve.

A closer look at Hubbert's curve reveals a critical feature that has great relevance for what might lie ahead. Hubbert observed that it took 110 years—from 1859 to 1969—to produce 227 billion barrels of cheap crude oil. Half of that oil was produced in the first 100 years. The second half was produced in less than ten years, from 1959 to 1969. Using the same model, Hubbert estimated in 1971 that the middle 80 percent of global oil production will be produced within fifty-eight to sixty-four years, or less than one lifetime.[51]

The Cassandras versus the Optimists

Geologists have combined Hubbert's curve with mathematical modeling to forecast the peak in global production. Campbell and Laherrère sparked the current debate over peak global oil production in a lengthy article published in *Scientific American* in March 1998. Campbell, who received his Ph.D. in geology from the University of Oxford, worked for Texaco as an exploration geologist and later at AMOCO as chief geologist for Ecuador. He later became exploration manager for AMOCO in Norway and executive vice president of FINA in Norway. Laherrère worked for Total, a French oil company, where he supervised exploratory techniques worldwide. His surveys led to the discovery of Africa's largest oil field. Campbell and Laherrère were, for many years, associated with Petroconsultants in Geneva, Switzerland, an oil-consulting firm that manages industry databases.

Campbell and Laherrère's study relied on a database kept by Petroconsultants that covers 18,000 oil fields worldwide. According to their data, there were only 850 Gbo of conventional oil in P^{50} reserves in the world in 1996. This figure is significantly lower than the 1,019 Gbo reported in the *Oil and Gas Journal* and the 1,160 Gbo figure published in *World Oil*. Allowing for the discovery of an additional 150 billion barrels of oil, Campbell and Laherrère estimate that the oil industry will be able to recover another 1,000 billion barrels of conventional oil. This represents slightly more oil than the 875 billion barrels that have already been produced.[52] Non-OPEC producers will peak in oil production before 2010, and the five leading OPEC producers in the Middle East—Saudi Arabia, Kuwait, Iraq, Iran, and Abu Dhabi—will likely peak around 2015.[53] Based on the collective data and computer modeling, the two geologists predict an overall peak in global oil production around the year 2010.[54]

A number of other geologists agree with Campbell and Laherrère's assessment. L. F. "Buzz" Ivanhoe is a petroleum geologist and former senior advisor of worldwide evaluations of petroleum basins for Occidental Petroleum. He believes that global oil supply will fail to meet global demand around 2010 and that supply will rapidly decline at about 3 percent per year thereafter.[55] When production peaks around 2010, says Ivanhoe, we should expect to see a spike in the price of crude oil and other fuel prices and, with it, global hyperinflation. He warns that a global crude oil shortage of just 5 percent "could bring back the gasoline lines of the 1970s . . . but this time the oil shortage will be permanent."[56]

Craig Hatfield, a geologist at the University of Toledo, using a base estimate of 1,550 Gbo remaining—a figure 55 percent higher than the one used by Campbell and Laherrère—nonetheless comes to a similar conclusion on the time line for peak global oil production. Hatfield starts with 1,000 billion barrels of reported global oil reserves—150 Gbo more than Campbell and Laherrère believe exist. He acknowledges that he may be too optimistic and that some of these reserves may simply be politically inspired calculations. He then goes on to estimate an additional 550 billion barrels of producible oil yet to be discovered. Again, significantly higher than Campbell and Laherrère's estimates. Combining the two figures, Hatfield theorizes that there are 1,550 billion barrels of oil yet to be produced. When this is added to the more than 800 billion barrels already produced, Hatfield comes up with 2,350 billion barrels of ultimately recoverable oil reserves.[57]

Assuming that the global rate of oil consumption continues to grow at 2 percent, Hatfield concludes that the world will have consumed half of the ultimate oil production before 2010. That date could be extended for several more years, he says, if the OPEC countries restrict oil production, as they have in the past, to keep prices high on world markets.[58]

James J. MacKenzie says that even if we assume that estimated ultimate recovery of oil is as much as 2,600 billion barrels, which is higher than most estimates, the peak for oil production would only be extended to the year 2019.[59]

Others have also weighed in on the debate. Franco Bernabe thinks that the global peak in oil production could come within the first decade of the 21st century, creating a 1970s-style oil crisis.[60] Kenneth S. Deffeyes, professor emeritus at Princeton University and formerly a pioneer petroleum engineer at the Shell Oil Research Laboratory in Houston, Texas, says that, depending on which of the Hubbert methods one uses—Deffeyes was a colleague of M. King Hubbert—the peak in global oil production could occur as early as 2003 and as late as 2009. Deffeyes bases his calculations on there being approximately 2.1 trillion barrels of recoverable crude oil reserves left.[61] He adds that

> no initiative put in place starting today can have a substantial effect on the peak production year. No Caspian Sea exploration, no drilling in the South China Sea, no SUV replacements, no renewable energy projects can be brought on at a sufficient rate to avoid a bidding war for the remaining oil.[62]

Deffeyes, who is clearly worried about the international consequences of an early peak in global oil production, quips, "at least, let's hope that the war is waged with cash instead of with nuclear warheads."[63]

The International Energy Agency of the Organization for Economic Cooperation and Development (OECD) estimates that world energy demand could grow by as much as 57 percent between now and 2020 and that global conventional-oil production will peak sometime in the second decade of the 21st century, between 2010 and 2020.[64]

The experts are divided, then, into two broad camps: those who believe that production of conventional oil is likely to peak in twenty-eight to thirty-eight years and those who think it is more likely to peak much sooner, within eight to eighteen years. Again, it should be emphasized that the optimists' and pessimists' projections of when global oil production is likely to peak differ by only ten to thirty years, which is a very small time window in history. While both camps believe that the age of cheap oil is coming to an end, the difference in time perspectives is critical to determining priorities, both in terms of energy policies and economic and political initiatives.

The optimists counter the new, more pessimistic computer-modeled studies with the argument that it is not true that no significant new oil fields remain to be discovered, and they point to the new giant field—the first in years—found off the coast of West Africa by the French company Elf, as well as to the discovery of two new super-giant fields in Kazakhstan and Iran. They are hopeful that other yet-to-be-found fields will provide an additional 5 billion barrels of oil each year, closing one-half of the gap between global supply and demand by 2010.[65] For example, the U.S. Geological Survey estimates that in the former Soviet Union there may be as much as 100 billion barrels of oil yet to be found. The U.S. Geological Survey is also optimistic in its projections for finding more oil in the Middle East and off the Atlantic shore areas of South Africa and South America.[66]

Colin Campbell claims that these figures are far too optimistic and says that, based on all the available data, the amount of ultimately recoverable conventional oil to be found in the former Soviet Union is only two-thirds of what the USGS projects.[67] As to the possibility of finding more giant fields like the one off the West African coast, virtually all of the critics concede that, whereas there may be a few more significant fields yet to be found, the likelihood of locating mega-fields of the size of those in Kuwait and Saudi Arabia is highly improbable. "There is only so much crude oil in the world," says Campbell, "and the industry has found about 90 percent of it."[68]

For the most part, the optimists pin their hopes more on new recovery technologies to extract more from existing fields than on the prospect of locating significant new fields. Douglas Bohi, an economist with Charles River Associates in Washington, D.C., believes, like others in the industry, that "the prospect is out there for an amazing increase in the [oil] reserve base."[69]

It is true that world oil reserves have risen steadily over the past two decades, and this has led geologists like William Fisher of the

University of Texas at Austin to conclude that "we're thirty, maybe even forty, years before the peak."[70] The USGS is convinced that the optimists are right and has revised upward the estimates of projected growth in ultimately recoverable reserves by 612 billion barrels.[71]

The question of growth of reserves in existing fields has as much to do with market conditions as with technological breakthroughs. If oil prices rise on world markets, development and utilization of new, more expensive drilling technologies become commercially viable options. Three new technologies are already adding significantly to growth in reserves. New 4-D seismic analysis allows geologists not only to track where oil, water, and gas are located in a field but also to predict where they will drill. The new monitoring and tracking technology can increase recovery by ten to fifteen percentage points in some fields, although the process can only be used in areas where soft rock is prevalent.[72]

Petroleum engineers have long known that conventional pumping of oil from wells, which continues until the oil slows to a trickle, can leave as much as 60 percent of the oil behind. To get that oil out, they pump natural gas, steam, or liquid carbon dioxide into depleted wells. The infusion spreads through the pores of the rock, pushing oil that would have been bypassed into adjacent wells. Water is also being pumped below the oil to increase pressure and force it to the surface. These techniques have increased the recovery factor by as much as 15 percent, although they are expensive. The new 4-D seismic monitoring adds 10 to 15 percent to the costs of producing a barrel of oil. The new injection techniques are even more expensive, increasing the cost of oil production by 50 to 100 percent.[73]

Directional drilling is still another technique for increasing recovery rates and is less expensive than injection. Engineers use sensing equipment that measures electrical resistance in surrounding rocks and, with new, more versatile drills, snake their way to the places where it is located.[74]

Optimists point out that in the 1960s only 30 percent of the oil in most fields was recovered. Now the recovery rate is 40 to 50 percent, and in a few years it could be as high as 75 percent in many fields.[75] The pessimists counter that "much of the technology is aimed to increase production rates . . . it doesn't do much for the reserves themselves."[76] Besides, monitoring, tracking, and drilling technologies have been developing continuously for the past 100 years and have been factored in all along the way in shaping the Hubbert curve. And when it comes to the biggest oil fields in the Middle East, the new technologies are of little advantage, since the oil needs no help in gushing up to the surface. The new recovery techniques will only result in small changes to the curve, say the pessimists. Albert Bartlett, a physicist and professor emeritus at the University of Colorado at Boulder, says that "all these things the economists talk about are just jiggling in a minor way with the curve." Bartlett's own forecast is that global oil production will peak as early as 2004.[77]

Finally, the optimists look to the oceans and argue that new oil fields likely exist 1,000 meters or more below the surface. New technologies allow engineers to locate and retrieve the oil, although the process is extremely expensive. Roger Anderson, Doherty Senior Scholar at the Larmont-Doherty Earth Observatory at Columbia University, writes enthusiastically of the new deepwater exploration in an article in *Scientific American*. Anderson and others believe that deepwater exploration will add 5 percent to the combined global reserves, but they concede that it won't make a decisive difference. "Although it is unlikely that these techniques will entirely eliminate the impending shortfall in the supply of crude oil," says Anderson, "they will buy critical time for making an orderly transition to a world fueled by other energy sources."[78]

At the end of the day, argue the pessimists, new discoveries and new recovery technologies are not going to significantly affect the timetable. Whether we are prepared or not, global production of conventional oil is likely to peak sometime between 2010 and 2020. Dr. Walter Youngquist, one of the deans of 20th-century geology, sums up the pessimists' case: "My observations in some seventy countries over about fifty years of travel and work tell me that we are clearly already over the cliff." Youngquist warns that "the momentum of population growth and resource consumption is so great that a collision course with disaster is inevitable."[79]

The Last Oil

While the optimists and pessimists disagree about when global production of conventional crude oil is likely to peak, they all acknowledge that most of the remaining reserves are in the Middle East and that it is only a matter of time before the world becomes dependent on the Persian Gulf for its growing oil needs. The U.S., long the leading supplier of oil, and until the 1950s responsible for more than half of world production, has experienced a steady decline in oil production since 1970—the year in which U.S. oil production peaked.[80] Since that time, the U.S. has had to import more and more oil. Today, the U.S. remains the largest user of crude oil. With only 5 percent of the world's population, the U.S. consumes nearly 26 percent of the world's oil. It only produces 11 percent of total world oil and currently holds only 2 percent of global reserves.[81] The EIA predicts that the U.S. will become even more dependent on foreign oil in the years to come, which is worrisome since oil imports figure prominently in the U.S. balance of trade deficit.[82]

Most Americans would no doubt be surprised to learn that we import a smaller percentage of oil from OPEC than we did twenty-five years ago. In the first six months of 2001, the U.S. bought more petroleum from Canada than from Saudi Arabia, and only two of our ten leading suppliers of oil were Middle Eastern nations.[83]

Over the next four to five years, the U.S. and other countries are expected to import an increasing percentage of their oil from Russia. Russian oil companies have invested billions of dollars in new exploration and drilling, and the Russian government has helped construct new pipelines to the Baltic and Black Seas. The result is that Russian oil production has increased to 7 million barrels per day in 2002, making it, at least temporarily, the largest oil producer in the world.[84]

In the fall of 2001, Russian oil flooded world markets, sending oil prices so low that OPEC openly threatened to retaliate with a price war if Russia did not curtail its output. OPEC threats notwithstanding, Russia is unlikely to significantly curtail its oil production. The Russian government, while no longer in a position to dictate terms to the country's privately owned oil companies, has a lot riding on Russian oil and gas exports. After all, $4 out of every $10 in government revenue come from energy exports.[85] The government's economic plans hinge, then, on maintaining oil exports at current or increasing levels and at a price of $20 per barrel, all of which creates a Catch-22 of sorts for the government. If Russian oil companies continue to flood world markets with oil, depressing prices to $17 or lower a barrel, the government could be stymied in its attempt to modernize the economy and improve the Russian standard of living. On the other hand, if oil exports are seriously constricted to keep prices high, the volume might not be sufficient to bring in enough oil revenue for the government treasury.

The Russian president, Vladimir V. Putin, is not unmindful of the new strategic position his country enjoys by dint of the country's oil exports. In October 2001, he told delegates to a world economic forum in Moscow that, as "instability in the world directly impacts world markets, Russia remains a reliable and predictable partner and supplier of oil."[86] Putin's veiled reference to tensions and instability in the Middle East, especially in light of the September 11th attacks on the World Trade Center, was not lost on the audience.

Russia's new status in the world oil market is likely to be a short-lived affair. The former Soviet Union's oil reserves have been steadily declining for two decades. In 1975, the USSR boasted reserves of 83 billion barrels. By the mid-1990s, the former Soviet Union—including all of the republics that previously were parts of the USSR—claimed reserves of less than 57 billion.[87]

Over the next few years, the topping-out of Russian oil production, as well as of oil from the North Sea, the Alaskan north slope, the areas off the shores of West Africa, and other regions, will leave the Middle East in the enviable position of supplier of last resort before the end of the decade. Even making allowances for inflated reserve figures, it is generally agreed that two-thirds of the present conventional-oil reserves in the world lie in the Middle East. Saudi Arabia alone possesses 26 percent of the total global reserves of oil.[88]

The real significance of Middle East oil becomes strikingly apparent when we look at the nature of the oil fields in the Persian Gulf. There are more than 40,000 known oil fields in the world, but only forty super-giant fields—those containing more than 5 billion barrels of oil—hold more than half of the oil reserves in the world. Twenty-six of those forty super-giant fields are in the Persian Gulf.[89] Moreover, while other giant fields, especially those in the U.S. and Russia, have peaked and are now on the decline, the Middle East fields are still ascending the bell curve. The reserve-to-production ratio (R/P) tells the story. The R/P is the number of years that reserves of oil will last at current production rates. In the United States, where more than 60 percent of the recoverable oil has already been produced, the R/P is 10/1. In Norway, the R/P is also 10/1, and in Canada it is 8/1. By contrast, in Iran the R/P is 53/1, in Saudi Arabia 55/1, in the United Arab Emirates 75/1, in Kuwait 116/1, and in Iraq 526/1.[90]

Oil production will gravitate back to the Middle East in the coming decade. According to the Energy Information Administration forecast, the Persian Gulf will provide an increasing amount of the world's production of oil in the years ahead.[91] Recall that at the time of the first oil crisis in 1973, the Middle East accounted for 38 percent of world oil production. In subsequent years, the Middle East's share of the world market fell to 18 percent as countries reduced their oil consumption and oil companies explored for oil in other regions of the world. With many of those other oil fields now being depleted, the Middle East share of the market has begun to climb again, and currently represents about 30 percent of the total market share. This time, however, says the EIA, the Middle East market share will continue to climb until it reaches half or more of global oil production.[92]

Industry analysts say that when the "swing producers"—the five key Middle East oil-producing countries—are able to control more than one-third of world production, they will be in a position once again to dictate the price of oil in world markets, as they did for a brief time in the 1970s. Much depends on when Russian crude exports begin to slow.

Campbell and his colleagues see a twofold process. In the first phase, the Middle East swing producers will command a significant share of world production—approximately one-third—allowing them to increase the price of oil. Ten years later—around 2015—the Persian Gulf countries will peak in oil production, sending oil prices through the roof.[93] Like it or not, says Walter Youngquist, "the Persian Gulf Muslim nations are geodestined to have the last word on oil."[94]

How high prices go will be determined by a number of variables. First, according to the IEA forecast, the five swing producers in the Middle East would need to increase oil production from 27 million to 48 million barrels per day between now and 2010 to meet increasing global oil demands.[95] However, current planned expansion in oil production in the Middle East is likely to fall short of what is needed by some 10 million barrels per day, leaving a dangerous shortfall in oil production that could significantly increase the price per barrel on world markets. Second, even the current expansion plans—which are too little to meet expected demand—will be so expensive that they will likely drive the price per barrel even higher, well before oil production peaks in the Middle East. Joseph Riva, formerly of the U.S. Congressional Research Service, warns that

> planned oil production expansions . . . are less than half that needed to meet the 2010 world oil demand projected by [the International Energy Agency], but will cost in excess of $100 billion, plus an additional $20 billion to upgrade and expand Persian Gulf refineries to meet growing world product demand. Oil production expansion beyond that planned would be even more expensive, on a per-barrel basis, as the remaining oil becomes more difficult to recover. . . .[96]

One way or another, say a growing number of geologists and industry analysts, the price of oil on global markets is going to go up, and probably much sooner than most people expect. The warning signs are everywhere. Still, as long as oil is relatively cheap and plentiful in the short run, few are willing to heed the storm clouds gathering on the horizon. The world is heading into troubled waters, say the Cassandras, and is unprepared for the consequences that lie ahead. And this time, the oil crisis will be not temporary but permanent, forcing a fundamental change in our way of life, with effects that will reach far into the future.

For many, the possibility that we may be running out of "cheap" available oil to power the industrial way of life is so unimaginable that any thought of it is likely to be regarded with incredulity, regardless of how many studies might be brought to bear on the subject. Our unconcern is understandable. Rarely do societies respond to an "anticipated" change in their circumstances. But when that potential change can radically affect the totality of our way of life and the very geopolitics of the world we live in, the collective nonchalance becomes a prescription for disaster.

While the prospect of a truly global energy crisis is a new phenomenon, human history is replete with examples of great civilizations that have failed to heed the warning signs, exhausted their regional energy regimes, and suffered catastrophic collapse. Making the right choices for the future requires an understanding of how past civilizations coped with their own energy crises. There are, in fact, inherent rules to how the energy game has to be played if civilizations are to survive, prosper, and continuously renew themselves. When those rules are ignored or given short shrift, societies grow old and die. Learning those rules is the most important education at hand as we ponder our own energy future.

3

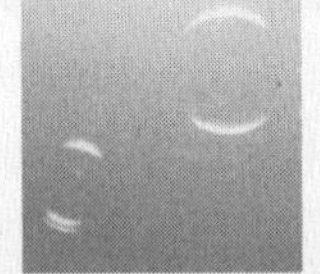

ENERGY AND THE RISE AND FALL OF CIVILIZATIONS

Frederick Soddy, the British Nobel laureate in chemistry, once noted that the indivisible currency upon which all of science is based is energy.[1] Each day, the sun's rays bathe the Earth with thousands of kilocalories of energy per square meter. Some of this energy is captured by living things and converted into forms useful to sustain life, while the rest ends up as heat and is radiated back into space.[2]

If energy is the alpha and omega of existence, then "power" is defined as "the rate of flow of useful energy."[3] All of life requires energy and sufficient power to maintain the rate of flow. The struggle for survival, then, both between and within species, is really a competition to capture useful energy and secure its continued flow through living systems.

Energizing Culture

Anthropologist Leslie A. White observes that, in the evolution of culture, human beings' first "power plant" was their own bodies. For most of human history, homo sapiens lived a hunter-gatherer existence and captured the energy stored in wild plants and animals. By acting collectively and cooperatively on their environments, they could increase their critical mass and use their human power plants to secure what they needed to sustain small kinship communities. Later, as our species made the transition from hunter-gatherers to pastoralists and farmers, human beings were able to sequester more energy from their environment. By domesticating animals and plants, they secured a continuous and reliable supply and surplus of readily available energy and, by so doing, increased the amount of energy that could flow through their bodies and communities. Plant cultivation—aided by irrigation systems—greatly increased the yield per unit of human energy or labor expended. Agricultural surpluses also freed at least some people from toil on the land. Freeing people from labor created the beginnings of a social hierarchy and the differentiation of tasks. Priest and warrior classes slowly emerged, as did an artisan class somewhat later on. The differentiation and specialization of tasks spawned new, more complex institutional arrangements, which in turn helped to facilitate an even greater energy flow-through.

The cultivation of cereals some 10,000 years ago in North Africa, the Middle East, China, and India marked a turning point for human society. The cereals have been called "the great moving power of civilization."[4] The food surpluses provided an energy endowment to sustain growing populations and the establishment of kingdoms, and later of empires. The great civilizations of Egypt and Mesopotamia rose in the wake of cereal cultivation. Large engineering projects were undertaken, including the establishment of elaborate hydraulic systems to irrigate fields. Women invented pottery, providing containers to store surplus grain for inventory and/or trade. The metallurgical arts aided in the development of more sophisticated weaponry for conquest and capture of additional land and slaves. Members of the non-producing priestly class used their time, in part, to track the movements of the planets and stars, giving them greater ability to predict spring floods and the best time to lay down seeds. Mathematics and writing also emerged in hand with the cereal civilizations. Mathematics provided the means to erect great monuments, most notably the pyramids of Egypt. Writing proved particularly helpful both in storing the collective knowledge of increasingly complex and diverse societies and in managing the communication flow of civilizations spread out over great distances.

The next shift, from agriculture to an industrial way of living, once again increased the amount of energy that could be captured, stored, and utilized—this time in the form of fossil fuels harnessed and processed by machines. The new machine energy acts as a substitute mechanical slave, multiplying the amount of energy and power available per capita and for the society as a whole.

George Grant MacCurdy, in *Human Origins*, writes of the human experience as an evolutionary journey in the increasing use of available energy: "The degree of civilization of any epoch, people, or group of peoples, is measured by ability to utilize energy for human advancement or needs."[5] Many anthropologists agree. White, for example, uses energy as a yardstick for measuring the success of all human culture. He argues that whether a culture is low or high in achievement is directly correlated to the amount of energy consumed per capita. The very function of culture, argue White and other anthropologists, is to "harness and control energy so that it may be put to work in man's service."[6] Human beings accomplish this by creating tools to capture and transform energy and communication mechanisms and social institutions to manage the process of energy transmission and distribution. What we call human progress, then, say MacCurdy and White, is to a large extent the wily ability of human beings to use symbolic forms, tools, and institutional arrangements to capture and use more and more energy and, by so doing, extend their power and increase their well-being. Howard Odum, one of the pioneers in the field of natural energy systems, cautions that in the coming together of "man, mind, and energy," one must bear in mind that it is the source of the energy, not human inspiration, that ultimately sets the limits on human progress. Odum writes:

> All progress is due to special power subsidies, and progress evaporates whenever and wherever they are removed. Knowledge and ingenuity are the means for applying power subsidies when they are available, and the development and retention of knowledge are also dependent on power delivery.[7]

The history of human civilizations—including their rise and fall—can't be properly understood unless we appreciate the importance of "power subsidies." The bottom line, for every society in history, says Odum, is the availability of energy surpluses. All the human creativity in the world will inevitably come up short in advancing the well-being of the race, absent sufficient energy reserves to be captured and harnessed.

White offers a shorthand for measuring the relationship between energy use and cultural evolution. There are, he says, three critical factors in assessing the "progress" of any culture: first, "the amount of energy harnessed per capita per year"; second, "the efficiency of the technological means with which energy is harnessed and put to work"; and third, "the magnitude of human need-serving goods and services produced."[8] Putting these factors together, White concludes that "culture evolves as the amount of energy harnessed per capita per year is increased, or as the efficiency of the instrumental means of putting the energy to work is increased."[9] White, very much in the materialist tradition of the European Enlightenment, is unequivocal in his belief that energy is the ruling factor, both in biological and cultural systems. "Thus we trace the development of culture from anthropoid levels to the present time," says White, "as a consequence of periodic increases in the amount of energy harnessed per capita per year effected by tapping new sources of power."[10]

It should be noted that MacCurdy, White, and Odum represent a particular school of thought. Certainly, all three could be faulted for equating increased energy use with social and cultural advances. An equally compelling case could be made that increased flow-through of energy in a society also correlates with greater coercion and subjugation of peoples and with environmental degradation.

Moreover, many anthropologists argue that, while the sequestering and transforming of increasing amounts of energy are germane, even essential, to the development of culture, that doesn't necessarily mean that culture itself—its symbols, tools, myths, and institutions—is to be understood solely as an instrumental means to advance the flow of increasing amounts of energy through the body politic. Still, putting aside, for the purposes of our discussion, the age-old argument of whether cultures serve some other end besides a purely materialist one, White, Odum, and MacCurdy's perspective is helpful for understanding how energy reservoirs and flows affect the evolution and devolution of cultures.

It is true that human societies have continued to increase the amount and quality of energy flowing through each individual life and the social life of society, at least since the dawn of the Neolithic revolution and the onset of an agricultural existence. The increasing flow, in turn, has required the use of more-sophisticated tools and more-complex institutional arrangements to both harness and manage the process. Advances in toolmaking and more complex institutional arrangements come with a heavy price tag, including the development of more hierarchical social structures, greater differentiation and specialization of human tasks, and more concentration of power at the top. In other words, the greater the horizontal flow of energy from environment to society, the greater the vertical flow of societal power from the top down needed to secure the process. This is not really surprising. Small hunter-gatherer bands relying on their own members as "power plants" generate only about ¹⁄₂₀ horsepower per capita per year.[11] That kind of energy flow-through did not require complex institutional arrangements to manage the process.

Later in history, our forebears captured and "harnessed" one another as energy-producing power plants—a process that continued apace until the latter part of the 19th century—and used human slavery as a means to increase energy flow. Slave labor built the great pyramids of Egypt, the Great Wall of China, and the ceremonial shrines of the Mayan and Teotihuacan civilizations in the Americas. The

Great Wall in China required the labor of more than a million slaves, half of whom perished in the effort.[12] Nearly 20 percent of the population of Rome in the first few centuries A.D. were slaves.[13] Human slaves were also used throughout history as "beasts of transport." Rickshaws still exist in great numbers in Asia, even to this day, although the labor is no longer forced. Galleys using slave labor to power them were still common as late as the 16th century in the Mediterranean.[14]

Slave societies required a far more elaborate use of tools, especially in the form of military armaments to capture slaves, as well as more advanced communication channels and transportation grids to efficiently commandeer, sequester, and mobilize slave labor. The result was greater concentration of power at the top of the social hierarchy to control and manage the increased complexity. But even that degree of concentration of power pales in contrast to the hierarchical controls needed in an advanced industrial society today to capture energy in the form of oil, coal, and natural gas and to manage the machine slaves that move energy through the society.

Energy is the elemental force and the medium upon which all human culture is built. And, yes, human history shows a marked increase in both the flow of energy and the complexity of social institutions needed to accommodate that flow. But to fully comprehend why civilizations built on different energy regimes rise and fall, we need to understand the rules that govern energy. These rules exist a priori as laws of nature and dictate the terms by which energy flows both here on Earth and in the universe at large. They provide the instructions for how the energy game has to be played to result in success for the player. These rules also tell us why past civilizations have failed and what our society needs to do to avoid a similar fate as we reach the "turning point" of the fossil-fuel era.

The Laws of Thermodynamics

Two laws govern all of energy. The first and second laws of thermodynamics weren't fully articulated until the latter half of the 19th century, nearly 170 years after Newton published his *Principia* outlining the laws of mechanics. Unfortunately, while every schoolchild is tutored in the laws governing gravity and why apples fall from trees the way they do, few are ever introduced to the laws governing the transformation of energy. Yet these laws are far more helpful in instructing us about the ebbs and flows of existence, including the passage of time and the workings of the Earth's chemical, biological, and social systems. Mathematician and philosopher Alfred North Whitehead once remarked to his students that Newtonian mechanics merely tell us about space-time relationships of matter in motion. Whitehead adds:

> As soon as you have settled . . . what you mean by a definite place in space-time, you can adequately state the relation of a particular material body to space-time by saying that it is just there, in that place: and, so far as simple location is concerned, there is nothing more to be said on the subject.[15]

The laws of thermodynamics, however, tell us how energy behaves and are therefore more relevant to understanding the day-to-day comings and goings of ecosystems and social systems. Albert Einstein once mused about which laws of science were the most inclusive and far-reaching. Turning to the laws of thermodynamics, the great 20th-century physicist opined:

> A theory is more impressive the greater is the simplicity of its premises, the more different are the kinds of things it relates and the more extended its range of applicability. Therefore, the deep impression which classical thermodynamics made on me. It is the only physical theory of universal content which I am convinced, that within the framework of applicability of its basic concepts, will never be overthrown.[16]

The first and second laws of thermodynamics state that "the total energy content of the universe is constant, and the total entropy is continually increasing."[17] The first law, that the total energy content of the universe is constant, is sometimes called the "conservation law." It means that energy can neither be created nor destroyed. The amount of energy in the whole universe has been fixed since the beginning of time and will remain so until the end of time. Every human being ever born and all the things that human beings have ever built over the course of history represent energy that has been transformed from one state to another. The energy that makes up the human body and the things we create started off somewhere else in a different state in nature before becoming embodied in a human or material form. When a human being dies and decomposes, and when a material thing disintegrates, the energy that is released finds its way back into nature.

This is where the second law of thermodynamics comes in. While energy cannot be created or destroyed, it is continually changing in form, but always in one direction, from available to unavailable. For example, if we burn a piece of coal, the energy remains but is transformed into sulfur dioxide, carbon dioxide, and other gases that then spread into space. No energy has been lost in the process, yet we can never re-burn that piece of coal again and get useful work out of it. The second law tells us that whenever energy is transformed, some amount of available energy is lost in the process; that is, it is no longer able to perform useful work. This loss of usable energy is referred to as "entropy," and it is one of the most important and least understood and appreciated concepts in physics. The term was coined by the German physicist Rudolf Clausius in 1868.

In order for energy to be converted into work, observed Clausius, there must be a difference in energy concentration (i.e., difference in temperature) in different parts of a system. Work occurs when energy moves from a higher level of concentration to a lower level (or a higher temperature to a lower temperature). For example, a steam engine does work because part of the system is very cold and the other is very hot. Equally important, when energy goes from one level to another, the result is that less energy is available to perform work the next time. Consider a red-hot poker: When it is removed from a fire, it begins to cool, because heat always flows from the hotter to the colder body. After a certain amount of time, however, the poker is the same temperature as the surrounding air. This is what physicists refer to as the "equilibrium state," the state where there is no longer any difference in energy levels and therefore no more ability to do work. The once-useful energy is no longer concentrated in the red-hot poker but now so randomly dispersed in the air that it is unavailable. The second law, then, tells us that energy is always transformed in one direction: from hot to cold, concentrated to dispersed, or ordered to disordered.

It is possible to reverse entropy, but only by using up additional energy in the process. And, of course, that additional energy, when used, increases the overall entropy. For example, recycling discarded wastes requires the expenditure of additional energy in the collecting, transporting, and processing of used materials, which increases the overall entropy in the environment. Let us say we take a chunk of metallic ore from the Earth's surface and fashion it into a utensil. Over the lifetime of that utensil, metal molecules are continually flying off the product as a result of friction and wear. The loose molecules are never destroyed, but they become so randomly dispersed that they are no longer available to do useful work. They eventually find their way back to the Earth, but in a much more widely scattered, dispersed state than when they were parts of a chunk of metallic ore. A device might be invented to collect and recycle all of the dispersed metal molecules scattered hither and yon, but both the energy used to run the machine and the machine itself would be using up available energy in the process—the first in the form of spent gases and the second in the form of molecules lost through wear and tear from friction. Again, to recycle the random metallic molecules, more available energy has to be used, increasing the overall entropy bill.

While the sun's energy is not expected to burn out for billions of years, and therefore will continue to bathe the Earth with new energy for as long as we can imagine, it must be understood that energy concentrated in material form here on Earth, whether it be metallic ores or fossil fuels, is relatively fixed in the geological time frame of importance to society. That is because, in thermodynamic terms, the Earth is a closed system in relation to the solar system and the universe. There are three kinds of thermodynamic systems: opened, closed, and isolated. Open systems exchange both energy and matter. Closed systems exchange energy but do not exchange any appreciable matter. Isolated systems exchange neither energy nor matter. The Earth is a closed system; that is, it exchanges energy with the solar system, but, except for an occasional meteorite or some cosmic dust, it does not exchange matter with the outside universe. The point that needs to be emphasized is that the solar inflow of energy does not by itself produce matter. The sun can flow into an empty jar forever without producing life of any kind. The Earth has a fixed endowment of terrestrial matter that can be converted into other useful forms, including life, with the help of solar energy. For example, the sun's energy interacted with terrestrial matter during the Jurassic era, helping transform that matter into life. That life, decomposed, makes up the carbon deposits that we burn today in the form of coal, oil, and natural gas. The spent energy, in the form of gases, is no longer available to do work. While similar carbon deposits might conceivably accumulate in some future period of geologic history, for all intents and purposes that time is so distant as to be irrelevant to human needs. That is why we call fossil fuels non-renewable sources of energy.

To summarize, the first law of thermodynamics states that all energy in the universe is constant; that it cannot be created or destroyed. Only its form can change. The second law states that energy can only be changed in one direction; that is, from usable to unusable, from available to unavailable, or from ordered to disordered. Everything in the universe, according to the second law, began as available concentrated energy and is being transformed over time to unavailable, dispersed energy. Entropy is the measure of the extent to which available energy in any subsystem of the universe is transformed into an unavailable form.

If society is organized around the continuous effort to convert available energy from the environment into used energy to sustain human existence, then Frederick Soddy's observation about the importance of the laws of thermodynamics seems apt. He argued that the laws of thermodynamics "control, in the last resort, the rise and fall of political systems, the freedom or bondage of nations, the movements of commerce and industry, the origins of wealth and poverty, and the general physical welfare of the race."[18]

But wait a moment. If, in fact, energy is continually being transformed from a concentrated to a random state or from an ordered to a disordered existence, how do we account for living things, and even social systems, which seem to maintain a high level of concentrated energy and order, in apparent defiance of the thermodynamic maxim?

Biologists, at first, were unsure how to reconcile life with the ironclad laws of thermodynamics. Harold Blum, in his book *Time's Arrow and Evolution*, helped put biology within the context of the first and second laws of thermodynamics by explaining that life is just a special case of how the laws of energy unfold. All living things, said Blum, live far away from equilibrium by continually absorbing free energy from the environment. While the organism maintains a steady state of orderly existence, it does so by feeding off available energy and increasing the overall entropy in the environment. Blum explains that "the small local decrease in entropy represented in the building of the organism is coupled with a much larger increase in the entropy of the universe."[19]

The source of free energy is the sun. Plants take up the sun's energy in photosynthesis and provide a source of concentrated energy that animals can then consume directly, by eating plants, or indirectly, by eating other animals. The Nobel laureate physicist Erwin Schrödinger observed that "by continually drawing from its environment negative entropy . . . what an organism feeds upon is negative entropy; it continues to suck orderliness from its environment."[20]

When we stop to think about it, what the biologists are saying makes common sense. We stay alive by continually processing energy through our bodies. If the energy flow were to stop, or our bodies were unable to process the energy properly because of disease, we would quickly move to death, or an equilibrium state. At death, the body begins to decompose rapidly, and our physical being dissipates and disperses into the surrounding environment. Life, therefore, say biologists, is an example of non-equilibrium thermodynamics; that is, life maintains order and remains far removed from equilibrium, or death, by continuous processing of free or available energy from the larger environment.

The process of maintaining a non-equilibrium state, away from death, is costly in energy terms. Even the plants, our most efficient "power plants" on Earth, absorb only a tiny fraction of the Earth's incoming energy through photosynthesis. The rest is dissipated.

Thus, the tiny entropy decrease in the plant is secured at the expense of a far greater entropy increase in the overall environment.

The 20th-century philosopher and mathematician Bertrand Russell noted that "every living thing is a sort of imperialist, seeking to transfer as much as possible of its environment into itself and its seed."[21] The more evolved the species is in nature's hierarchy, the more energy is required to maintain it in a non-equilibrium state and the more entropy is created in the process of keeping it alive.

Consider the case of a simple food chain consisting of grass, grasshoppers, frogs, trout, and humans. According to the first law, energy is never lost. But, according to the second law, available energy is turned into unavailable energy at each step of the food-chain process, increasing the overall entropy in the environment. Chemist G. Tyler Miller reminds us that, in the process of devouring the prey, "about 80 percent to 90 percent of the energy is simply wasted and lost as heat to the environment."[22] In other words, only 10 to 20 percent of the energy of the prey is absorbed by the predator. That is because the act of transforming energy from one life form to another requires the expenditure of energy and therefore results in the loss of energy. The amount of free energy needed to keep each more evolved species up the food chain alive is staggering. Miller calculates that "three hundred trout are required to support one man for a year. The trout, in turn, must consume 90,000 frogs, which must consume 27 million grasshoppers, which live off of 1,000 tons of grass."[23] Every living thing up the evolutionary ladder, then, maintains itself in a non-equilibrium, ordered state at the expense of creating greater disorder (dissipated energy) in the overall environment.

Energy is continually flowing through every living organism, entering the system at a high level and leaving the system in a more degraded state in the form of waste. Again, the more evolved the organism, the more energy it requires to sustain itself against equilibrium. This means that as we travel up the evolutionary ladder, each succeeding species has to be better equipped physiologically to capture available energy. Biologist Alfred Lotka says that we can think of living things as "transformers" of energy. "The close association of the principal sense organs: eyes, ears, nose, tastebuds, tactile papillas of the fingertips with the anterior [head] end of the body, the mouth end, all point to the same lesson," says Lotka.[24] Natural selection favors those organisms that are able to "increase the total mass of the system, rate of circulation of mass through the system, and the total energy flux through the system . . . so long as there is present an unutilized residue of matter and available energy."[25]

Rethinking Economic Progress

Evolution, then, is a building up of more complex systems of organization, with each species more differentiated and specialized than the last in order to capture and concentrate more available energy. Seen from the point of view of thermodynamics, evolution is less a matter of steady, uninterrupted advance than a continuous tradeoff between increasing uses of energy and increasing dissipation of energy. Evolution results in the creation of larger islands of order at the expense of the creation of even greater seas of disorder in the world. If this is true for species and ecosystems, it is equally the case for human social systems. Lest there be any doubt on this score, consider how much free energy is required to sustain the economic and social structures and lifestyles of Americans and how much entropy is created in the process.

The United States has created a massive complex of "transformers" to move energy through every artery of the social organism. Although, as mentioned, the United States is home to less than 5 percent of the world's population, its people consume about 25 percent of the world's total produced energy.[26] The average American uses 8,000 pounds of oil, 4,700 pounds of natural gas, 5,150 pounds of coal, and one-tenth of a pound of uranium each year.[27]

Geologist Walter Youngquist says that if we want to understand exactly how much energy is flowing through American society each day, we can do so by calculating how much additional equivalent "person-power" each individual has at his or her disposal. Start with the assumption that one "person-power" (PP) = .25 horsepower = 186 watts = 635 Btu/hr. If, says Youngquist, the present energy use in the U.S. were calculated in terms of the equivalent person-power it would take to provide us with the same amount of work, it would require nearly three times as many people as presently exist in the world.[28] In day-to-day terms, the average American's energy diet is equivalent to having fifty-eight energy slaves working continuously, twenty-four hours a day.[29] If "we purchased the energy in a barrel of oil at the same price we pay for human labor ($5/hr), it would cost us over $45,000," rather than the current cost of $25 a barrel.[30]

Not surprisingly, the entropy bill for maintaining an American energy diet far from equilibrium is equally high. While Americans consume 25 percent of the world's energy, they also contribute 30 percent of the world's carbon dioxide emissions. About 6.6 tons of greenhouse gases are emitted per person every year in the United States. Eighty-two percent of those emissions are caused by the burning of fossil fuels to generate electricity and provide power to run our automobiles, buses, trucks, and planes. Most of the remaining emissions are from methane produced in landfills and by modern agriculture practices as well as by natural-gas pipelines and industrial chemicals.[31]

It is also the case that as society matures and ages, it requires more and more energy just to be maintained, which means that less energy is available for innovation and expansion. The American Society of Civil Engineers' (ASCE) 2001 report on the state of America's infrastructure is revealing in this regard. According to the association, America's infrastructure—its roads, bridges, transit, aviation, schools, drinking water, wastewater, dams, solid waste, hazardous waste, navigable waterways, and energy—is in such disrepair that an investment of $1.3 trillion will be required over the next five years to fix it. The ASCE reports that 33 percent of the nation's major roads are substandard, costing American drivers more than $5.8 billion a year. Poor roads also contribute to 13,800 highway fatalities each year. Moreover, one-third of the country's urban freeways are now congested. The nation's bridges don't fare any better. Twenty-nine percent of the bridges are structurally deficient or obsolete, and the association says it will take $10.6 billion per year for the next twenty years to fix them. The nation's airports are overcongested and their facilities are stressed, resulting in as many as 50,000 flight delays per month. America's schools are aging and severely overcrowded, and the association estimates that 75 percent of their infrastructures are inadequate to meet educational needs. The 54,000 drinking-water systems are antiquated and becoming a source of non-point pollution. Many of the country's 16,000 wastewater systems are near collapse. Some of the sewer systems are more than 100 years old and buckling from the demands placed on them. There is currently a $12 billion annual shortfall in the funding of the nation's wastewater infrastructure. More than 2,100 dams are now classified as unsafe, and sixty-one dam failures were reported in 1999 and 2000. The inland waterway systems are so old that 44 percent of all the lock chambers have already exceeded their fifty-year design lives. Yet transportation demands on America's waterways are expected to double by 2020. Finally, since 1990, the nation's electrical capacity has experienced an annual shortfall of 30 percent. Currently, 7,000 megawatts (MW) of electricity are being added each year, far short of the 10,000 megawatts per year required to meet the 1.8 percent annual growth in demand.[32]

So, the more evolved and complex the social organism, the more energy is required to sustain it and the more entropy is produced in the process of maintaining it. This simple reality flies in the face of orthodox economic theory. In fact, neither capitalism nor socialism is capable of accommodating the harsh "real-world" realities imposed on society and the environment by the first and second laws of thermodynamics.

Classical capitalist theory is wedded to the idea that economic activity turns waste into value. John Locke, the English Enlightenment philosopher, argued that "land that is left wholly to nature . . . is called, as indeed it is, waste."[33] Locke turned the second law of thermodynamics on its head by proclaiming that nature itself is useless and only becomes of value when human beings apply their labor to it, transforming it into productive assets. Locke writes:

> He who appropriates land to himself by his labour, does not lessen but increases the common stock of mankind. For the provisions serving to the support of human life, produced by one acre of inclosed and cultivated land, are . . . ten times more than

> those which are yielded by an acre of land, of an equal richness lying waste in common. And therefore he that incloses land and has a greater plenty of the conveniences of life from ten acres than he could have from a hundred left to nature, may truly be said to give ninety acres to mankind.[34]

Capitalist economics is steeped in the earlier physics of Newtonian mechanics and has never come to grips with the laws of thermodynamics. Borrowing from Newton's idea that for every action there is an equal and opposite reaction, classical economists such as Adam Smith and Jean-Baptiste Say likened the marketplace to a mechanism in which supply and demand are continually readjusting to each other. If consumers' demand for a good or service goes up, sellers will raise their price accordingly. If the sellers' price becomes too high, demand will slacken, forcing the seller to lower the price, to rekindle demand. The same logic was applied to the harnessing of natural resources. If resources become scarce, the seller's price will rise, encouraging suppliers either to use new technologies to tap more hard-to-get supplies or to find alternatives to those resources. The overall resource base is considered inexhaustible and always available, in some form, for the right price. The entropy bill, on the other hand, is considered, if at all, as an externality of doing business and marginal to the overall costs of conducting commerce.

The laws of thermodynamics tell us something quite different. Economic activity is merely borrowing low-entropy energy inputs from the environment and transforming them into temporary products and services of value. In the transformation process, more energy is expended and lost to the environment than is embedded in the particular good or service being produced. Even the "finished" product or service is only temporary, and upon use or consumption it dissipates or disintegrates, eventually ending up back in the environment in the form of used-up energy or waste.

What, then, are we to conclude about the nature of Gross Domestic Product (GDP)? We think of GDP as a measure of the wealth that a country generates each year, whereas, from a thermodynamic point of view, it is more a measure of the temporary energy value embedded in the goods or services produced at the expense of the diminution of the available energy reserves and an accumulation of entropic waste. If even the goods and services we produce eventually become part of the entropy stream, then for all of our notions of economic progress the economic ledger will always end up in the red. That is, when all is said and done, every civilization inevitably ends up sucking more order out of the surrounding environment than it ever creates and leaves the Earth more impoverished because it has existed.

The societies that last longest are the ones that create as near a balance between nature's budget and human social budgets as possible, given the inevitable restraints imposed by the second law. So-called "steady state" societies learn to live, as well as they can, with nature's own timetable. The harnessing, converting, distributing, and consuming of energy, in its many economic forms, are maintained at a rate roughly commensurate with the surrounding environment's ability to recycle wastes and restore renewable energy resources. The ratio can never be one-to-one because of the inherent energy penalty built into the transformation process. Still, some societies, especially hunter-gatherers and small kinship-based agricultural communities, have maintained themselves for long periods of time before exhausting their energy regimes. The great civilizations in human history have been less successful. Re-looking at the rise and fall of the world's great civilizations through the lens of thermodynamics can throw much-needed light on the current crisis facing our civilization as we approach the watershed for an energy regime based on the exploitation of fossil fuels.

Why Great Civilizations Collapse

Civilizations are a rare phenomenon in world history. The British historian Arnold Toynbee says that there have only been thirty. Other historians are less generous. Oswald Spengler, A. L. Kroeber, Carroll Quigley, and Rushton Coulborn argue that there have been between eight and fifteen great civilizations worthy of the name.[35]

What we do know is that about 6,000 years ago a new type of social organization began to emerge in various parts of the world. Small kinship-based communities began to give way to new entities that bear many of the markings of states. Tribal affiliations, the traditional organizing framework for social life, were subsumed into large territorial-based affiliations. Diverse peoples came together, their new boundaries defined by region and geography. These states brought with them hierarchical control exercised by powerful ruling elites. Economic activity was marshaled and overseen by a centralized government. Bureaucracies were created to manage the daily affairs of the subjects, legal codes were established to regulate human behavior, statutes were enacted and mechanisms were put into place to collect tribute in the form of taxes on agricultural production. Armies were conscripted to plunder neighboring lands, to ensure security against invaders, and to maintain internal order. Power was lodged in urban centers and emanated out to distant borders of the kingdom. These centers were sacred sites and thus had divine legitimacy. The leaders, in turn, were designated as "emissaries of the gods," responsible for administrating the earthly kingdoms. Increasing agricultural surpluses and plunder of foreign lands, as well as slave labor, provided the available energy reserves to maintain a growing urban population of non-producers and to sustain the complex, far-flung social arrangements. Civilizations are distinguished from simpler societies by the marshaling, processing, and consuming of large amounts of energy.

Two questions that have plagued historians from ancient Greece to modern times are: Why are there so few civilizations (they appear to be an anomaly in the long human sojourn)? And why does the power they amass and institutionalize, while seemingly impregnable, often disintegrate quickly and lead to sudden collapse?

Historians have offered up a number of interesting explanations for the rise and fall of great civilizations. The German historian Oswald Spengler favors an organic model, preferring to view the birth, life, and death of a civilization much like the cycles that govern human life itself. Every civilization, observes Spengler, "passes through the age-phases of the individual man. Each has its childhood, youth, manhood and old age."[36] Like every individual human being, argued Spengler, each civilization lives by "an idea," by which he meant a concept of its own unique identity, passion, and feelings, and a sense of its mission and destiny. Most important, notes Spengler, a civilization's very "living existence, that sequence of great epochs which define and display the stages of fulfillment, is an inner passionate struggle to maintain the Idea against the powers of Chaos . . ."[37] "The Idea," for Spengler, consists of the organizing principles and schema that marshal the collective energy of the people and create an ordered social universe. His thesis that every civilization organizes its human energy around a central idea to create order "against the powers of Chaos" puts him squarely in the thermodynamic camp. He is not alone. Arnold Toynbee takes a different approach to dissecting history and explaining the rise and fall of civilizations, but in the end he comes to an analysis that also resonates with the laws of thermodynamics. Toynbee sees the development of civilization as a series of challenges and responses. Societies confront the problems of how to marshal resources and harness the energy that those resources make available to them. Challenges that threaten to impede the energy flow arise continuously, requiring the society to respond in creative new ways. Toynbee believes that the collapse of great civilizations occurs because of a "failure of vitality," an inability to mobilize human energy sufficient to overcome the obstacles that threaten to undermine the workings of the society.[38]

In ancient times, religious renewal provided the catalytic agent to unify and mobilize human energy to the task of continually remaking society. When the religious spirit wanes, human energy dissipates. The collective faith that got people to combine their energy and act as one, to realize a common mission, disintegrates. Human behavior becomes more random. The collective will becomes fragmented and prey to the forces of chaos and despair. The historian Rushton Coulborn notes that societies remain vigorous as long as their religious commitment is strong, and that they become weak when their religious zeal diminishes.[39] In modern times, ideology has played a similar role in mobilizing the collective energies of a people. Again, the laws of energy lurk in the shadows, a constant reminder that the rise and fall of civilizations, like the birth, life, and death of every living creature, comport to the stringent rules of non-equilibrium thermodynamics.

Joseph A. Tainter, in his seminal work, *The Collapse of Complex*

Societies, provides a useful framework for understanding the dynamics of why civilizations collapse. His theory of "marginal returns" would be familiar to every engineer who has ever had to contend with energy flows and entropy in machines.

Tainter agrees with Leslie White, Frederick Soddy, Bertrand Russell, and others that human history has been characterized by the creation of increasingly complex technological and social arrangements for the capturing of free available energy from the environment. Greater energy flow-through, in turn, allows human settlements to grow. Populations increase, social life becomes more dense and varied, and culture advances. Societies collapse when the energy flow is suddenly impeded. Energy is no longer available in sufficient volume to sustain the increased populations, defend the state against intruders, and maintain the internal infrastructure. Collapse is characterized by a reduction in food surpluses; a winnowing of government inventories; a reduction in energy consumed per capita; disrepair of critical infrastructures like irrigation systems, roads, and aqueducts; increasing popular defiance toward the state; growing lawlessness; a breakdown in central authority; a depopulation of urban areas; and increasing invasions and pillaging by marauding groups or armies.

Collapse sets in, according to Tainter, when a mature civilization reaches the point at which it is forced to spend more and more of its energy reserves simply maintaining its complex social arrangements while experiencing diminishing returns in the energy enjoyed per capita. For example, in the youthful period of a civilization's history, energy reserves are used to raise armies and make armaments for the purpose of conquest of other lands and people. The plunder—in terms of human slave labor, confiscated land, and treasure—brings in more energy than is expended in securing it. In the latter stages of civilization, the state often spends more of its energy reserves defending existing territory against invasion and plunder from the outside, resulting in greater energy expenditures, with little or no new energy returns.

Tainter makes the same point with agricultural production. In the early stages, irrigation systems are created, land is cleared, and roads are built to move grain from country to town. The energy expenditures result in a net increase in energy returns. In the later phase of a civilization's history, the state spends more money just maintaining the existing agricultural support infrastructure as well as the bureaucracies of state that oversee the society. Much of the available energy also goes toward sustaining the lifestyles of the power elites or other non-productive members of society. To accommodate greater expenditures of energy for maintenance, the land is often overworked to secure additional increments of energy, leading to soil degradation, erosion, and diminishing yields. A large population, whose numbers grew during the good times, suddenly enjoys less energy per capita even though the people are working longer and harder. At the same time, the state often imposes more taxes on its subjects to make ends meet, further hastening the downward spiral. To forestall collapse, the society uses up much of its remaining energy reserves and excess capacity in a desperate effort to maintain itself in its previously existing non-equilibrium state, bringing it ever closer to a "turning point." Societal unrest, in turn, forces the rulers to spend whatever energy reserves remain on maintaining a semblance of law and order, leaving even less energy flowing through the system to the people. Oftentimes, at the end-stages of a civilization, what food surpluses are left, as well as the valuable energy resources, are commandeered by the state to feed and equip the military itself, further angering the public and creating a kind of vicious positive feedback. The population begins to disaggregate and fend for itself, setting off the process of disintegration. Unless a new energy subsidy is found—either by conquest or the exploitation of a new energy source—collapse is all but inevitable.[40]

The Thermodynamics of Rome

The Roman Empire offers a good case study in the politics of energy. In many ways, Imperial Rome is far closer to the modern world in its way of life and in its economic, social, and political organization than to the ancient world that gave birth to it. The lessons of its rise and fall are still fresh enough in the historical record to provide guidance for us as we wrestle with our own future.

Rome owed its greatness to its brilliant military conquests.

Roman armies captured Macedonia in 167 B.C., annexing the land and seizing the wealth of its king. The royal Roman treasury became so large that it allowed the Romans to stop taxing their own citizens. Shortly thereafter, Rome annexed the kingdom of Pergamon, and the bounty doubled the Roman state budget overnight. The conquests of Syria, in 63 B.C., and then of Gaul brought gold and untold wealth to the empire.[41] The military conquests were so successful, from an economic point of view, that they paid for themselves and provided surpluses for continued military ventures. Slave labor, mineral resources, forestland, and cropland provided an ever-accelerating flow of available energy for the empire. The expansionary period ended with Augustus's conquest of Egypt. So much wealth accrued from the annexation that Augustus celebrated by distributing coins to the plebians of Rome.[42]

After suffering a series of military defeats to the Germans and others, Rome retrenched and devoted its energy to building an infrastructure to maintain its empire. The shift from a conquest-based regime to a colonizing regime proved costly. With no new revenues coming in from newly captured lands, Rome found itself without sufficient funds to provide for basic services. Augustus instituted a 5 percent tax on inheritances and legacies to pay for the retirement of military personnel. The imposition of the tax angered the Romans, who had not been taxed at all during the period of the late Republic.[43]

The expense of maintaining a military was particularly burdensome. The standing army drained energy from the empire and took away gains previously enjoyed by the Roman population. The cost of salaries, rations, billeting of troops, and equipment continued to go up. So too did the cost of maintaining public works and a bloated civil service. The public dole had also grown considerably during the heyday of expansion and now had to be maintained with declining revenues. During the reign of Julius Caesar, nearly one-third of the citizens of Rome were receiving some form of public assistance. Between the years A.D. 41 and 54 alone, more than 200,000 families received free wheat from the government.[44]

The sheer logistics of maintaining a vast empire became increasingly costly. Garrisoning troops throughout the Mediterranean and

Europe, maintaining roads, and administering annexed territories consumed more and more energy, while the net return in energy secured from the territories steadily dropped. Marginal returns had set in. In some instances, it cost Rome more money to maintain certain colonies—Spain and England, for example—than revenue generated.[45]

No longer able to maintain its empire by new conquests and plunder, Rome was forced to look to the only other energy regime available to it: agriculture. The story of Rome's gradual decline is intricately bound up with the waning fortunes of its agriculture production.

The popular conception is that Rome collapsed because of the decadence of its ruling class, the corruption of its leaders, the exploitation of its servants and slaves, and the superior military tactics of invading barbarian hordes. While there is merit to this argument, the deeper cause of Rome's collapse lies in the declining fertility of its soil and the decrease in agricultural yields. Its agricultural production could not provide sufficient energy to maintain Rome's infrastructure and the welfare of its citizens. The exhaustion of Rome's only available energy regime is a cautionary tale for our own civilization as we begin to exhaust the cheap, available fossil fuels that have kept our industrial society afloat.

Italy was densely forested at the beginning of Roman rule. By the end of the Roman Imperium, Italy and much of the Mediterranean territories had been stripped of forest cover. The timber was sold on the open market and the soil converted to crops and pastureland. The cleared soil was rich in minerals and nutrients and, at first, provided significant production yields. Unfortunately, the denudation of forest cover left the soil exposed to the elements. Wind blew across the barren landscapes, and water ran down from the mountaintops and slopes, taking the soil with them. Over-grazing of livestock resulted in further deterioration of the soil.

The continuing decline of soil fertility came just at the time when Imperial Rome began to rely on its agriculture to provide an energy substitute for its failing foreign conquests. During the later period of the Roman Empire, agriculture provided more than 90 percent of the government's revenue.[46] Food production had become the critical linchpin in the survival of Rome.

Supporting a growing urban population of non-producers put increasing strain on small farms. Agricultural production intensified to keep up with the food requirements of the urban population and the army. The overworked land further diminished soil fertility, which in turn led to even more intense exploitation of an already exhausted soil base. Small farmholders could not produce sufficient yield on the eroded soil to pay the yearly taxes (the government imposed a fixed tax on the land irrespective of yield). Farmers borrowed to stay in business. Much of the debt they incurred went to pay taxes, and little or none was left over with which to make improvements. Without any other source of revenue, farmers could not afford to let their soil lie fallow long enough to be rejuvenated. Instead, they continued to lay down seed on an ever-more-exhausted soil base, which left them with even lower yields and greater debt loads. Small farmers were forced to sell their holdings or give them up to the lending institutions. Throughout Italy and the Mediterranean, large landowners bought up smallholdings, creating great landed estates—the latifundia. Much of the land was no longer fit for growing food and had to be converted to pasture to raise livestock. Impoverished and uprooted, Italian farmers migrated to the cities, where they went on the public dole. In the fourth century, more than 300,000 people in the city of Rome were receiving some form of public aid.[47] The growth in city expenditures to subsidize the lifestyles of the wealthy; provide for the welfare of the poor; finance public works and government bureaucracies; pay for the erection of monuments, public buildings, and amphitheaters; as well as the costs involved in financing public games and displays all stressed the agricultural-energy regime beyond its limits. Depopulation of the countryside took place throughout the empire. In some of the provinces of North Africa and throughout the Mediterranean, up to one-half of the arable land had been abandoned by the third century.[48]

Depopulation of the countryside also had other repercussions. Abandoned land was no longer being stewarded. The result was even greater erosion and loss of soil fertility. Lowlands were hit particularly hard by the massive depopulation of the countryside. Wet fields were no longer drained in the early spring and therefore were left swampy.

The spreading swampland became a breeding ground for mosquitoes, which caused a spread of malaria. The disease weakened an already dispirited and hungry population, further sapping the human energy reserves.[49]

Plagues broke out in the second and third centuries, killing up to one-third of the population in some regions of Italy.[50] The reduction in population meant fewer people available to farm and to fill the ranks of the civil service and military. The situation became so desperate that the government was forced to reintroduce conscription to raise its army. In 313, the emperor Constantine issued an edict requiring some sons of soldiers to be automatically conscripted—establishing a hereditary military service.[51] Similar statutes were enacted in the early fourth century, creating a hereditary civil service.[52] More controversially, around the same time, Constantine instituted the colonate, which in effect bound agricultural workers to the land they presently lived on—establishing the concept of serfdom.[53] This practice would outlive the Roman Empire and remain in place until the great Enclosure Acts were enacted in the 1500s in Tudor England, freeing serfs from bondage to the land.

The colonate was too little too late. By the fourth century, the rural population was so diminished that the remaining numbers, even when yoked by law to the land, were insufficient to rev up agricultural production in dying fields.

Rome was experiencing the harsh realities imposed by the laws of thermodynamics. Maintaining its infrastructure and population in a non-equilibrium state required large amounts of energy. Its energy regime, however, was becoming exhausted. With no other alternative sources of energy available, Rome put even more pressure on its dwindling energy legacy. By the fifth century, the size of the government and military bureaucracy had doubled.[54] To pay for it, taxes were increased, further impoverishing the population, especially the dwindling farm population. The empire, writes Joseph Tainter, began consuming its own capital in the form of "producing lands and peasant populations."[55]

Weakened by a depleted energy regime, the empire began to crumble. Basic services dwindled. The immense Roman infrastructure fell into disrepair. The military could no longer hold marauding invaders at bay. Barbarian hordes began to whittle away at the decaying Roman Empire, at first in distant lands. By the sixth century, the invaders were at the gates of Rome. The great Roman Empire had collapsed. By the sixth century, the population of Rome, once numbering more than 1,000,000, had shrunk to less than 30,000 inhabitants. The city itself was reduced nearly to rubble, a stark reminder of how unforgiving the energy laws are.[56]

The entropy bill was enormous. The available free energy of the Mediterranean, North Africa, and large parts of Continental Europe, reaching as far north as Spain and England, had been sucked into the Roman machine. Deforested land, eroded soil, and impoverished and diseased human populations lay scattered across the empire. Europe would not recover for another 600 years.

LIKE ROME, the industrial nations have now created a vast and complex technological and institutional infrastructure to sequester and harness energy. The global industrial economy relies almost exclusively on fossil fuels to maintain itself in a highly ordered non-equilibrium state—nuclear power and renewable sources of energy account for a small percentage of the energy mix in the industrialized countries. So dependent is every aspect of our current economic and social systems on cheap crude oil and natural gas that, when they do become increasingly scarce to find, process, and use, we risk a potentially destabilizing series of breakdowns in systems and subsystems across the spectrum of modern life. Could such breakdowns have a positive feedback, a so-called cascading effect, as was experienced in Rome, with each failure feeding off the other, resulting in systems spiraling out of control and our way of life unraveling? Could developments proceed so quickly that we might not have the time necessary to put in place a new energy regime of sufficient magnitude to ensure our future?

Understanding the architecture of our current fossil-fuel infrastructure should now be our most important priority. It is critical that we be able to identify the many structural weaknesses that exist all along the scaffolding we have built to move fossil-fuel energy through our industrial system. That scaffolding includes not only the technological infrastructure but also the economic, social, and political infrastructures that are built on top of it.

Even the most ardent supporters of fossil-fuel energy doubt that the hydrocarbon era will survive the 21st century. Whether we are able to reinvent our civilization along new energy lines, or instead suffer debilitating assaults on the current energy infrastructure, will depend on how willing we are to hold up our current energy regime to a rigorous re-evaluation and review. If we are to successfully take ourselves to where we need to go, we need to know exactly how we got to where we are now. The critical question to ponder is, what went wrong—and why—in how we harnessed energy in the Industrial Age? Finding the answer is crucial to ensure that we don't repeat the same mistakes in the creation of a new energy regime for the 21st and 22nd centuries.

4

THE FOSSIL-FUEL ERA

Remove fossil fuels from the human equation and modern industrial civilization would cease to exist. We heat our homes and businesses with fossil fuels, run our factories with fossil fuels, power our transportation with fossil fuels, light our cities and communicate over distances with electricity derived from fossil fuels, grow our food with the help of fossil fuels, construct our buildings with materials made from fossil fuels, treat sickness with pharmaceuticals made from fossil-fuel derivatives, store our surpluses with plastic containers and packaging made from fossil fuels, and produce our clothes and home appliances with petrochemicals. Virtually every aspect of modern existence is made from, powered with, or affected by fossil fuels.

Early in the 20th century, oil surpassed coal in the U.S. and other highly industrialized nations as the most important fuel in the fossil-fuel energy mix. Motor vehicles use the lion's share of oil consumption—about one-third of all the global oil consumed each year. There are currently 520 million automobiles in the world.[1] The United States accounts for 132 million of them. The U.S. also boasts 1.9 million trucks, 715,000 buses, and 21,000 locomotives. There are 11,000 large commercial aircraft, 28,070 ships, and 1.2 million fishing boats in the world, all powered by oil.[2]

Industry is the second-largest user of oil in the United States, at 23 percent of the total. More than one-fourth of the industrial oil is used as chemical feedstocks.[3] Petrochemicals are used to make thousands of products—from television parts to pharmaceuticals. Six percent of the oil is used for residential and commercial heating, and 4 percent of the oil is used to generate electricity in American power plants.[4]

Oil is among the most versatile of all the substances found in nature. One barrel of oil can produce

> enough gasoline to drive a medium-sized car over 200 miles; enough distillate fuel to drive a large truck almost 40 miles . . . enough liquefied gases to fill 12 small (14.1 ounce) cylinders for home, camping, or workshop use; nearly 70 kilowatt hours of electricity at a power plant . . . asphalt to make about one gallon of tar; about four pounds of charcoal briquets; wax for 170 small birthday candles, or 27 wax crayons [and] lubricants to make about a quart of motor oil.[5]

The transformation to a fossil-fuel civilization occurred more quickly than any other change in energy regimes in world history. Just 130 years ago, three-quarters of the total fuel supply in the U.S. was in the form of wood. Wood was used not only for heating but also as fuel for steamboats and railroads. Much of industry at the time was still powered by windmills and watermills.[6] In 1890, fewer than 10 million tons of oil were being produced in the world, mostly in the form of kerosene for illumination. At the beginning of the 20th century, oil still made up less than 4 percent of the world's energy. By the time of the Arab oil crisis in the 1970s, 2,500 million tons of oil were being consumed annually—a 200-fold increase in seventy years.[7] Today, more than 85 percent of the world's energy comes from fossil fuels—40 percent from oil, 22 percent from coal, and 23 percent from natural gas. Nuclear and hydroelectric power account for an additional 7 percent each, while only 1 percent comes from geothermal, solar, wind, wood, and waste sources. World energy use has increased seventy-fold since the onset of the fossil-fuel era.[8]

How History Is Really Made

Schoolchildren are taught that a young genius named James Watt gave birth to the fossil-fuel era and the Industrial Revolution with the invention of a steam engine powered by coal. The steam engine, in turn, ushered in a modern era of material progress and human achievements unrivaled in history.

Well, not exactly. The story of how the world turned to coal and steam power, opening the door to the fossil-fuel era, is a bit more convoluted.

Medieval Europe had long relied on wood as its primary energy source. The dense forest cover across western and northern Europe provided a seemingly inexhaustible source of fuel. By the 14th century, however, wood was becoming increasingly scarce. New agricultural advances, including new drainage technologies, the cross plow, the introduction of three-field rotation, and the use of horse teams for plowing, helped boost the amount of land under cultivation and dramatically increased food production. Food surpluses led to an increase in the human population, which, in turn, put more pressure on farmers to overexploit existing farmland as well as deforest more marginal lands for cultivation. By the 14th century, Europe was facing an entropy problem not dissimilar to the one that faced Rome in the second, third, and fourth centuries. A growing population was using up its energy resources faster than nature could replenish them. Widespread deforestation and eroded soil created an energy crisis. Historian William McNeill writes:

> Many parts of Northwestern Europe had achieved a kind of saturation with humankind by the 14th century. The great frontier boom that began about 900 led to a replication of manors and fields across the face of the land until, at least in the most densely inhabited regions, scant forests remained. Since woodlands were vital for fuel and as a source of building materials, mounting shortages created severe problems for human occupancy.[9]

The depletion of wood was as serious a problem for late medieval society as the depletion of oil is for us today. Like oil, wood was an all-purpose energy resource, used for a hundred and one different things. Historian Lewis Mumford lists some of the many ways wood was used at the time.

> The carpenters' tools were of wood but for the last cutting edge: the rake, the oxyoke, the cart, the wagon, were of wood; so was the wash tub in the bathhouse; so was the bucket and so was the broom; so in certain parts of Europe was the poor man's shoes. Wood served the farmer and the textile worker: the loom and the spinning wheel, the oil press and the wine presses, and even a hundred years after the printing press was invented, it was still made of wood. The very pipes that carried water in the cities were often treetrunks: so were the cylinders of pumps . . . the ships of course were made of wood and . . . the principal machines of industry were likewise made of wood.[10]

Mumford sums up the overriding importance of wood as an energy regime for medieval life, observing that "as raw material, as tool, as machine, as utensil and utility, as fuel and as final product wood was the dominant industrial resource."[11]

Much of the deforestation throughout the 15th century was undertaken to make room for expanding agricultural fields. In the 16th and 17th centuries, even more trees were felled to supply wood ash for cottage industries, including glassworks and soap. In England, the biggest strain on forests came from the increasing needs of the British navy. The production of iron and the building of ships required vast amounts of wood. Repeated attempts to regulate the cutting of forests proved futile. By 1630, wood had become two and one-half times as expensive as it had been in the late 15th century.[12]

Slowly, coal became a substitute for wood—first in England and later on the continent. A new energy regime began to take hold. But it should be pointed out that the switch to coal was not exactly greeted with unrequited joy. Quite the contrary. Coal was considered an inferior energy resource: hard to mine, transport, and store; dirty to handle, and polluting when burned. Edmund Howes lamented that "inhabitants in general are constrained to make their fires of sea-coal or pit coal, even in the chambers of honorable personages."[13] Still, by 1700, coal had begun to replace wood as the primary energy resource for England. By the mid-19th century, most of Europe had begun to convert to coal as well.[14]

Mining coal was a difficult affair. After exhausting the more readily available supply near the surface, miners had to descend deeper into the ground. At a certain depth, water levels were reached, and drainage became a serious obstacle to bringing up the coal to the surface. In 1698, Thomas Savery patented the first steam pump, providing miners with a tool to bring the water to the surface and thus remove coal from greater depths.

Coal was also much heavier and more bulky to transport than wood. It could not be easily moved by horse-drawn wagons over unsurfaced roads. The weight of the wagons, especially during rainy weather along rutted muddy roads, made transport all but impossible. Horse teams were also becoming increasingly expensive to use to pull the loads. With land for cultivation becoming more scarce, it was simply too costly to graze horses on precious agricultural soil. The answer to the transport problem came in the form of the steam locomotive set on iron wheels. The steam locomotive became one of the first energy machines of the fossil-fuel age and the forerunner of a new era.

We are used to thinking of material progress as a steady, uninterrupted flow of new and better ideas substituting for older, more primitive ways of doing things. In fact, human advance is more of a trial-and-error process and is often motivated by desperation. "Necessity," it turns out, is far more likely to be "the mother of invention."

And, as the conversion from wood to coal demonstrates, the switch in energy regimes is often regarded, in the early stages, as onerous and unwelcome. That is because human beings always seek out the easiest available energy resources to exploit first. As long as our hunter-gatherer ancestors had available to them an abundant supply of energy in the form of wild plants and animals, they had no need to switch over to a more arduous agricultural way of life. Similarly, forests are a far more accessible source of energy to harness, transform, and use than is coal. Surveying the history of human civilization from an energy perspective, historian Richard Wilkinson makes this rather telling observation:

> During the course of economic development man has been forced over and over again to change the resources he depended on and the methods he used to exploit them. Slowly he has had to involve himself in more and more complicated processing and production techniques as he has changed from the more easily exploitable resources to the less easily exploitable. . . . In its broadest ecological context, economic development is the development of more intensive ways of exploiting the natural environment.[15]

Coal, for example, is a more accessible form of available energy and easier to harness than oil and natural gas. That is why, as countries move from more easily available energy resources to forms of energy that are harder to find and process, the technological, economic, and social infrastructures, by necessity, become more complex, hierarchical, and centralized.

Our oil-based civilization is the most hierarchical and centralized energy transformer in all of history. We have created a complex non-equilibrium social organism with oil flowing through every artery. While industrial life has been hugely successful for those who are its immediate beneficiaries, it's fair to say that the very complexity that made it possible now threatens its very existence. That is because the social organism, like any living organism, operates as a whole. Every subsystem of this highly organized industrial civilization is totally dependent on the continuous flow of nonrenewable energy—in the form of oil and, to lesser extents, of coal and natural gas—throughout its internal environment, just as a living organism requires the continuous flow of blood cells through its body. If the flow of oil slows, the whole organism weakens.

The oil business has long been associated with the romantic lore of swashbuckling prospectors wildcatting for oil and unscrupulous entrepreneurs lurking in the wings waiting to take advantage of their good luck. That all happened. However, the oil saga is also about the step-by-step assemblage of a monumentally complex, highly fragile energy regime, so centralized in its command and control structure that it is more vulnerable to disruption or even sudden collapse than any previous energy infrastructure.

Midwifing the Birth of Oil

The first thing to bear in mind about oil is that it is unevenly distributed around the world. The United States became the world's leading industrial power in the 20th century, in large measure because of its rich oil deposits. Similarly, England's success at the beginning of the Industrial Revolution was, to a large extent, attributable to the vast domestic coal deposits at its disposal.

Although we no longer think of the U.S. as an oil power, we need to remember that, for the first seventy years of the 20th century, Texas oil wells were as much a symbol of America's greatness as was the Ford assembly line. Access to vital energy resources has always been a critical factor in the success of nations. While it is difficult for us in the West to comprehend, nearly half of the human race—more than 2.5 billion people—still rely on wood, animal manure, and crop residue for their fuels.[16]

On August 27, 1859, Edwin Laurentine Drake, a retired railroad conductor and misfit who referred to himself as "the Colonel," using a makeshift cable drilling rig, struck oil at a depth of 69.5 feet near the small town of Titusville, Pennsylvania. The oil flowed up to the surface at a rate of twenty barrels a day.[17] The oil age had begun.

The early oil industry was helped along by the Civil War. Crude oil was used to provide lubricants for machinery, illumination for lamps, and turpentine substitutes for the army. (Oil was not yet used as a fuel for locomotives.) By the end of the Civil War in 1865, oil was being pumped from the ground in West Virginia, New York, Ohio, and as far west as Colorado and California. There were fifty-eight oil refineries in Pittsburgh, thirty in Cleveland, and more scattered around the East Coast, mostly making kerosene for lamp oil.[18]

In 1868, a former clerk and bookkeeper from Cleveland, John D. Rockefeller, founded the Standard Oil Company of Pennsylvania. Rockefeller realized that the key to success in the oil business lay not just at the wellhead but also in owning the refineries and in controlling the transportation and marketing of the finished products. He went about the task of entering into favored arrangements with the railroads and later bought up pipelines. By 1879, the Standard Oil Company controlled nearly 95 percent of the refining capacity in the country.[19] Rockefeller soon began to buy up oil fields as well, with the dream of integrating every aspect of the oil business under one roof. His operations expanded so fast that by 1882 he had amalgamated his vast holdings with the establishment of the Standard Oil Company of New Jersey. The trust owned shares of dozens of industrial companies. By 1906, Rockefeller's power over the flow of oil in the U.S. had become so great that the government brought suit under the Sherman Antitrust Act. In 1911, the U.S. Supreme Court ordered the breakup of the Standard Oil Group and the divestiture of all shares held by the group in subsidiary companies.[20] The legal action was of little long-term consequence. After the holding company was dissolved, the individual companies incorporated within states where they were doing business. Many of the same men who held shares in the trust continued to hold shares in the various companies. Eventually, millions of shareholders came to own the many Standard Oil offshoots.

Meanwhile, other oil companies formed along the same lines as Rockefeller's, each a fully integrated petroleum enterprise, owning oil fields, pipelines, and refineries and controlling the transport and marketing of products all the way to the local gas station. By the 1930s, the major oil companies that would come to define the biggest industry in the world were in place. They included Standard Oil of New Jersey, Gulf Oil, Humble, Atlantic Refining Company, Sinclair, Standard Oil of Indiana, Phillips 66, Sucony, Sun, Union 76, and Texaco. Together, twenty-six companies owned two-thirds of the capital structure of the industry, 60 percent of the drilling, 90 percent of the pipelines, 70 percent of the refinery operations, and 80 percent of the marketing.[21]

Two developments in the opening years of the 20th century propelled oil to the center of American life and made America, in turn, the most powerful country in the world. The first was the invention of the internal combustion engine and the second the critical role that oil would come to play in securing victory for the U.S. and its allies in the two world wars.

The New Mobility

Karl Benz and Gottlieb Daimler of Germany were the first men to successfully put an internal combustion engine on wheels. Their horseless carriage debuted in 1885.[22] The engine was fueled by gasoline derived from crude oil. Although the engine was a German invention, it was Henry Ford's mass-production assembly line, churning out millions of affordable gasoline-powered vehicles, that made oil and automobiles the linchpin of a new age.

The first U.S. filling station opened for business in 1911 in Detroit.[23] The dizzying speed of automobile production caught the oil business by surprise. In an effort to keep pace with the almost insatiable demand for gasoline, energy companies expanded their exploration, opening up new fields almost weekly. By 1916, 3.4 million autos were on the U.S. roads. Just fourteen years later, there were more than 23.1 million cars in the U.S.[24]

Automobiles became the centerpiece of industrial capitalism for the rest of the twentieth century. Many other critical industries were tied to the fortunes of the car. Autos consumed "20 percent of the steel, 12 percent of the aluminum, 10 percent of the copper, 51 percent of the lead, 95 percent of the nickel, 35 percent of the zinc, and 60 percent of the rubber used in the U.S."[25] Industrialists gushed over the great commercial possibilities opened up by the automobile. One analyst, writing in 1932, observed:

> Think of the results to the industrial world of putting on the market a product that doubles the malleable iron consumption, triples the plate glass consumption, and quadruples the use of rubber! . . . As a consumer of raw material, the automobile has no equal in the history of the modern world.[26]

The automobile put millions of people on the road. It also brought town and country together and spawned the suburban culture while undermining traditional notions of neighborhood and community. And, more than any other invention of the 20th century, the automobile quickened the pace of life, making speed and efficiency the paramount virtues of our time.

Automobile production was responsible for much of the incredible economic growth experienced by the U.S. in the first three decades of the 20th century, and by Europe and Asia after World War II, but it was oil that made it all possible. The British statesman Ernest Bevin once remarked that "the kingdom of heaven may be run on righteousness, but the kingdom of earth runs on oil."[27]

While the automobile made oil indispensable to 20th century commercial and social life, it was the two world wars that convinced policy leaders of oil's strategic importance in the affairs of nations. Britain had long relied on its superior naval power to maintain its position of world dominance. In the first decade of the new century, the German government launched an ambitious effort to overtake the British navy on the high seas. A young British politician, Winston Churchill, was made First Lord of the Admiralty in 1911 and was charged with the responsibility of meeting the German threat. Churchill was mindful of the stakes in maintaining British naval superiority. He once exclaimed: "The whole fortunes of our race and Empire, the whole treasure accumulated during so many centuries of sacrifice and achievement, would perish and be swept utterly away if our naval supremacy were to be impaired."[28]

Churchill became convinced that the key to victory over Germany in any future war rested with converting the British navy from coal to oil-fired power. Oil-powered ships would be faster, would require fewer men to operate the engine rooms, would allow for greater radius of action, and could be refueled, if necessary, at sea. Churchill convinced the British government to begin making a switch to an oil-powered navy. In 1912, the British government agreed to build the first five oil-fired battleships.[29] That single decision would prove fateful in securing the Allies' victory over Germany a few years later during World War I.

Once committed to an oil-powered navy, Churchill and other British military commanders turned their attention to the question of ensuring a steady, uninterrupted supply of oil. At the time, the Royal Dutch Shell Company dominated the oil market in Europe. Worried that the company might fall under the influence of Germany, Churchill convinced the British government to invest in and become part-owners of a British energy company, Anglo-Persian. In addition, the government secretly negotiated a deal with the company to provide the admiralty with a twenty-year supply of oil.[30]

The gasoline-powered internal combustion engine played a significant role in securing victory over land, as well as on the sea, during the war. The British military invented an armored vehicle powered by an internal combustion engine, and placed it on tracks. The new vehicle, later referred to as "the tank," was able to cross barbed-wire fences and roll over heavily manned trenches, changing the balance of power on the battlefield.

World War I also saw the introduction of motorcycles, jeeps, and trucks. Quicker and easier to maintain than horses and more flexible than locomotives in delivering troops and supplies to the front lines, the new vehicles gave the Allies a critical edge over German forces.

The gasoline-powered airplane was used by both sides for reconnaissance and as a weapon in World War I. By the end of the war, more than 200,000 airplanes had been built and sent into the skies, changing the very concept of warfare. From now on, war could be fought from the sky as well as on land.[31]

Just a few days after the signing of the Armistice in 1918, British officials, meeting at a dinner for the Inter-Allied Petroleum conference in London, reflected on the significance of oil in the war effort. Lord Curzon, the chairman, proclaimed that "the Allied cause had floated to victory upon a wave of oil."[32]

Access to oil became even more important during World War II. Indeed, the entire strategy of the war revolved around controlling vital oil supplies. Soon after coming to power, Adolf Hitler turned his attention to securing uninterrupted oil supplies to power a reemergent German war machine. Hitler envisioned the army of the future relying less on sheer brute manpower and more on mobility and lightning speed of armored transport. The Blitzkrieg would become the moniker for German military action in the field. Without oil, Hitler's vision would be impossible. The problem was that Germany, while rich in coal resources, had no oil to speak of. Hitler realized that his country's dependency on coal-fired military equipment had cost them World War I and was anxious not to repeat the mistake of the past.

The German government set its sights on a two-pronged approach to securing oil: first, developing a synthetic-fuel industry at home; and second, pushing into a war with Russia to secure access to the rich oil fields at Baku in central Asia.

By the late 1930s, oil had already surpassed coal as the chief energy source in the U.S. In Germany, however, coal was still supplying 90 percent of the country's energy on the eve of World War II. Earlier in the century, German chemists had successfully extracted liquid from coal, creating synthetic fuel, but the process was expensive and couldn't compete with the cheap cost of crude oil on world markets. Hitler remained undeterred. In 1936, he launched an ambitious domestic synthetic-fuel industry with the help of German chemical giant IG Farber, proclaiming that "this task must be handled and executed with the same determination as the waging of a war, since on its solution depends the future conduct of the war."[33] In 1940, German refineries were producing 72,000 barrels of synthetic oil a day, nearly 46 percent of Germany's total oil supply. By 1944, the synthetic-fuel industry was supplying 57 percent of the total energy for the military effort, including 92 percent of the aviation fuel.[34]

Securing the rest of its oil needs proved even more difficult and costly. Desperate for oil, Germany invaded the Soviet Union on June 22, 1941. Hitler hoped that a quick victory over the Russians would give him access to the Baku oil fields in the Caucasus and provide sufficient fuel to win the war. Albert Speer, the German minister of armaments and war production, admitted, during interrogation by the Allies in 1945, that "the need for oil certainly was a prime motive" in the German decision to invade Russia.[35]

Russian resistance, however, stalled the German offensive. In August 1942, the German army reached the oil fields of Maikop in the Caucasus, only to find that the Russians had already blown up the wells and the refinery facilities. Short on fuel and far from home, the Germans were unable to defeat the Russians and secure Grozny, the oil heartland of the Caucasus.[36] The irony, notes Daniel Yergen, is that "the Germans ran short of oil in their quest for oil."[37] Thwarted in their efforts, the German government increasingly fell back on its domestic synthetic-oil production, which was not sufficient to maintain the military effort.

The Japanese attack on Pearl Harbor in 1941, like the German attack on Russia earlier in the year, was motivated by the desperate quest to secure access to oil for its military machine. The United States and the Dutch East Indies—now called Indonesia—were the main suppliers of oil to Japan. When the Japanese government invaded southern Indo-China in July 1941, the British, the Dutch Indies, and the Americans immediately placed an embargo on all exports of oil to Japan. With oil supplies running low, Japan made the decision to mount a surprise attack on the American fleet at Pearl Harbor. The objective was to cripple or destroy the U.S. Pacific Fleet and then to seize the oil fields of the Dutch East Indies. Their plan succeeded. But as the war progressed, the U.S. government was able to achieve air superiority over the Pacific, and, by 1944, American planes and ships were sinking Japanese tankers faster than they could be built. Oil shipments to Japan dropped by 50 percent in 1944 and to nearly zero in 1945. With no oil left to run their military operations, the Japanese were left to power their planes with fuel made from the roots of pine trees, and they even used charcoal to power their jeeps.[38] The outcome of World War II boiled down to the simple fact that the Allies controlled 86 percent of the world's oil supply.[39]

The Oil Empire

When one thinks of the global oil industry, the first term that is likely to come to mind is "Big Oil." The petroleum industry is the largest business in the world and is valued at between $2 trillion and $5 trillion.[40] The industry is made up of a vast complex that includes oil fields, off-shore oil platforms, thousands of miles of pipelines, giant oil tankers, refineries, computerized systems to manage the flow of fuel to end users, and filling stations, as well as thousands of companies producing petrochemical products ranging from lubricants and fertilizers to plastics and medicines. For most countries, oil is the largest item in their balance of payments.[41] Three of the seven largest publicly traded companies in the world are energy companies. Exxon Mobil is ranked number two in the Fortune 500, with revenues of $213 billion.[42]

The years 1999 and 2000 saw the creation of what industry analysts call "super major" energy companies. BP merged with Amoco and ARCO, Exxon merged with Mobil, Total merged with Elf, and Chevron merged with Texaco. The $200 billion in industry mergers is transforming big oil into colossal oil. The consolidation taking place among publicly traded energy companies puts them on the same playing field with giant state-owned oil companies, including Saudi Aramco, Petroleos de Venezuela, Iran's NIOC, and Mexico's Pemex. The former group controls much of the downstream business. Exxon/Mobil, Royal Dutch/Shell, BP, and Total Fina Elf now control 32 percent of global market sales and 19 percent of refining capacity. The latter group controls much of the upstream business. Saudi Aramco, Petroleos de Venezuela, NIOC, and Pemex produce 25 percent of the world's oil and hold 42 percent of the reserves. Together, just ten to twelve super majors and state oil companies dominate world energy.[43]

In the United States, five companies, Exxon/Mobil, Chevron-Texaco, BP Amoco-ARCO, Phillips-Tosco, and Marathon, control 41 percent of domestic oil exploration and production, nearly 47 percent of domestic refining and 61 percent of the domestic retail market.[44] After tax profits for the five companies rose from $16 billion in 1999 to $40 billion in 2000, a 146 percent increase in twelve months. After-tax profits for the same companies in the first quarter of 2001 rose again, from $8.7 billion to $12 billion, a 38 percent rise in just three months.[45] The oil industry's soaring profits stand in stark contrast to the 43 percent decline in income of the 1,400 other largest U.S. corporations in the first quarter of 2001.[46] Interestingly, American companies cite increased labor and "fuel costs" as mostly to blame for falling profit margins.[47] Because so few companies now control energy flow through the economy, they are in a unique position to dictate the terms of doing business for all of the other enterprises that make up the industrial mix. In a domestic and global economy that is increasingly controlled by fewer and fewer big corporate players in every industry and field, the energy companies are ensconced on the very top of the global pyramid, dispersing the energy that keeps every other economic activity going.

Finding, extracting, transporting, refining, and distributing oil and petrochemical-based products is a costly and complicated business. Only the largest companies in the world—those with deep pockets—have the financial wherewithal to manage the process from the wellhead to the gas pump. Consider just the activation index, the "measure of the total investment required to establish access to new oil, expressed in dollars per barrel per day of stabilized production."[48] In 1999, Iraq announced that it wanted to triple its oil production from two million to six million barrels per day. The cost of financing a tripling in oil production would be around $30 billion, or a development cost of $7,500 per barrel of oil per day of new production.[49] The world average activation index in 1999 was $2,000 per barrel of oil per day. In some places, like the deep waters in the Gulf of Mexico, the activation costs can be as much as $9,000 per barrel per day. The total cost in capital investment in worldwide exploration and production of oil is expected to run upward of $1 trillion over the next ten years.[50] Big companies like Exxon/Mobil, BP, and Shell each enjoys cash flows many times greater than in Saudi Arabia. The world's leading energy companies routinely invest $1 billion or more in new mega-projects like deep-water platforms.[51] In 1999, the eight largest oil companies financed more than 80 percent of the research-and-development spending in new exploration and production.[52]

Imagine, if you will, how demanding and complex the technology and support infrastructure must be to process, transform, and move millions of barrels of oil each day across the world and into every corner of daily life without significant delays or disruptions in the flow line. There are countless systems and subsystems, and they all have to be coordinated to ensure a steady stream of energy throughput. Geologist Robert O. Anderson likens this process to a great flowing "Amazon of oil."[53]

The energy industry uses more varied technology and specialized personnel than any other industry. Exploration requires the use of satellites and a knowledge of geology, geochemistry, and geophysics. Sophisticated computers and software programs are used to collect three-dimensional reflection seismic data and to create 3-D seismic images of the Earth's interior. Drilling wells to depths that can reach 20,000 feet require complex, high-technology oil equipment. Drilling platforms on the high seas are a modern engineering marvel, built to withstand hurricanes and typhoons. Analyzing the well bore and rock cuttings, what geologists call "logging the well," involves the use of neutron and gamma-ray instruments to record the resistivity curves at the well site. Mud engineers must be brought in to lubricate the drill bits with special drilling mud. Their job is to fit the drilling mud to the conditions in the well, making sure that the mud keeps the hole from caving in during drilling and when the cuttings flow to the surface. Laying hundreds of miles of pipelines across some of the most forbidding terrain in the world and maintaining them require specialized engineers as well. The refining process itself is perhaps the most complicated stage. Organic chemists break down the crude oil hydrocarbon complex and reconstitute it into a range of products from gasoline to plastics.[54]

Coordinating the entire flow process—that is, moving the oil through the various channels and transformations on the way to end users—brings into play a second group of management personnel and marketing specialists. R. O. Anderson walks us through the Byzantine maze of activities that are the oil business.

To begin with, 60 percent of the oil produced is already committed, often to the producers' own refineries. The rest is put up for sale on the open market. The buyers and brokers, sometimes representing other major oil firms or acting as independents, then resell the crude oil to buyers for other smaller refineries. The buyer must be well versed in the intricacies of the oil business, including knowing which refineries can process heavy, solid, sweet, or light crudes. He or she must also have a working knowledge of which petroleum products are needed in a given geographic region and which crude is best suited to produce those products.

The refiner, in turn, must be careful to distinguish the unique properties of each crude. The crude from Saudi Arabia, for example, has different properties than the crude from Venezuela. Oil companies identify crude by gravity, viscosity, wax content, and sulfur content. Today's refineries are built to handle particular feedstocks. For example, if a refinery were to run a high-sulfur crude through a system designed to handle low-sulfur materials, the machinery would corrode, with damages running into the hundreds of millions of dollars.

The marketing system can be equally vexing. Many petroleum-based product sales vary by season. Heating-oil sales rise in winter, while gasoline sales are usually higher in the summer. Management forecasts future needs six months to a year in advance, making sure that the right crudes are being channeled into the system and routed to the appropriate refineries, which in turn churn out the products required for each season. If a particular refinery finds itself with too much crude, because of a slack in demand or a temporary shutdown of the facility for repair, it will trade its oil with one or more refineries whose needs and requirements match, among the several hundred other refineries operating in the United States. Anderson makes the point that the oil industry is itself a pipeline and designed to flow. If oil is held in storage tanks, ships, or pipelines for too long, money is lost. Moreover, because it is a flow system, storage capacity is limited at every point in the process. Most operators maintain a working inventory of less than fourteen days, which means that the coordination process for ensuring that the oil flows to exactly where it is needed has to be well honed.[55]

Energy-company marketing departments, says Anderson, are divided into industrial, wholesale, and retail sectors and by specialty product sales. The largest industrial users, including airlines, utilities, chemical plants, and refineries, generally contract directly with the energy companies. Industrial sales run the gamut from asphalt to aviation fuels, coke for the metal and rubber industries, and natural-gas liquids for chemicals and agricultural products. Gasoline and automotive production still make up 50 percent of petroleum sales in the United States.[56]

Restructuring Commerce

The oil-energy infrastructure is far and away the most complex energy grid ever created. The nature of the energy has dictated the terms of its engagement. Unevenly distributed, difficult to extract, costly to transport, complicated to refine, and multifaceted in the forms in which it is used, oil, from the very beginning, has required a highly centralized command-and-control structure to finance exploration and production and coordinate the flow of oil to downstream end users. The highly centralized oil infrastructure inevitably gave rise to commercial enterprises organized along similar lines. In the discussion of the emergence of industrial capitalism, little attention has been paid to the fact that the energy regime determined, to a great extent, the nature of the commercial forms that took shape.

In a society built to rely on wood energy, the commercial enterprises are small and local, and the products are generally traded within limited geographic markets. Firms are most often family owned and require little in the way of outside financing, as the capital equipment is unsophisticated and relatively easy to assemble with the knowledge, tools, and other resources readily available in the community. Wood, as a power source, is limited. The pace, flow, and volume of production made possible by wood energy are not so great as to result in a qualitative change in both the speed and diversity of commercial activity, which would necessitate more coordination and more centralized and hierarchical command-and-control mechanisms.

Fossil fuels are of a different ilk. Coal, oil, and natural gas are more concentrated forms of energy, and, when properly harnessed, they speed the throughput and density of economic activity. The new pace and interactivity spill over into both the political and cultural arenas as well, forcing the rise of centralized, hierarchical command-and-control mechanisms across every sector to manage the increasing weight of human relationships.

The railroad and telegraph, among the first modern industries to rely on fossil fuels as a power source, established the operational framework for the new, more centralized and hierarchical enterprises that would come to dominate every industry and become the hallmark of 20th-century capitalism.

The railroad era began in earnest in the United States in the 1850s. In that single decade, 21,000 miles of rail were laid down.[57] The change in the speed of transport was without parallel in history. For the first time, human beings could eclipse the upper limits of speed imposed by their own legs, animal transport, and the harnessing of the Earth's winds and waves. Where a trip from New York City to Chicago in 1850 took three weeks or more, by 1857, trains were shuttling passengers between the two cities in only seventy-two hours.[58] The railroad allowed commercial enterprises to ship their goods quickly, cheaply, and reliably. They could also transport three times the load that canal service could carry for the equivalent resource costs.[59]

Keeping track of freight on thousands of railcars along thousands of miles of track, maintaining rail beds, keeping engines and cars repaired, overseeing thousands of employees, and ensuring safe passage required a new kind of organizational model. Economic historian Alfred Chandler points out that railroads were the first modern business enterprise to separate ownership from management. No single family could underwrite the huge capital costs of building a railroad. So, in the 1850s, European investment houses began financing U.S. railroads.[60] The operation of the railroads was put under the stewardship of a new class of professional managers. They created the precursor of the modern corporation, a centralized, hierarchical command-and-control mechanism made up of critical decision makers at the top of the pyramid and middle managers at various rungs who were responsible for specific day-to-day functions of the organization and accountable to higher-ups for overall guidance.

By the 1890s, says Chandler, the railroads' organizational structure looked remarkably like the one that would come to dominate 20th-century business. At the top of the organizational chart was the board of directors, and directly beneath were the president, general superintendent, and treasurer. Below the officers of the company were general managers, who supervised two or three general superintendents. On the next level down were middle managers responsible for specific departments and functions, including the head of machinery; the head of rail-bed maintenance, freight, and passenger traffic; the controller; the head of the legal department; the chief engineer; and the head of construction.[61] "The railroads," says Chandler, were "the first modern business enterprises."

> They were the first to require a large number of salaried managers; the first to have a central office operated by middle managers and commanded by top managers who reported to a board of directors. They were the first American business enterprise to build a large internal organizational structure with carefully defined lines of responsibility, authority and communication between the central office, departmental headquarters and field units; and they were the first to develop financial and statistical flows to control and evaluate the work of the many managers.[62]

By 1891, the Pennsylvania Railroad employed 110,000 workers, which made it the largest commercial enterprise in the world at that time. The company even began to rival the federal government in size. In 1893, the U.S. government's expenditures were $387.5 million. That same year, the Pennsylvania Railroad reported expenditures of nearly $100 million, making it the second-largest revenue-generating institution in America.[63]

The revolution in rail transportation went hand in hand with a revolution in communications. The telegraph allowed people to communicate instantaneously over great geographic distances for the first time. The railroads were the first to take advantage of the telegraph, using it to route traffic, track shipments, and coordinate right-of-way for trains coming from both directions on single tracks. By 1866, one company, Western Union, dominated the telegraph business. It would be synonymous with the term "telegraph" from that time on.[64] The first American communications giant used much the same type of organizational structure as did the railroads.

Both the economies of scale and speed dictated the terms of development of American railroads and the telegraph. Large amounts of capital were required to build and maintain the operations, and highly centralized command-and-control functions were necessary to coordinate the increased pace, flow, speed, and density of commercial activity. Economists, long used to thinking of markets as made up of small, independent sellers and buyers coming together to make rather simple exchanges of goods and services, soon began to talk instead about the merits of "natural monopoly."

The railroad and the telegraph, besides providing the new organizational model for doing business, also provided the core infrastructure for the advent of the modern factory system. Quick, dependable, year-round transportation and instantaneous communications allowed companies uninterrupted access to suppliers upstream and retail markets downstream. Industry, which had long been seasonal, could now operate 365 days of the year. Coal, and then oil, provided energy to illuminate and heat factories and to power equipment. The capital costs of maintaining a fossil-fuel energy grid favored large factories over small shops. Large factories, in turn, required centralized command-and-control mechanisms to coordinate activity.

The modern business bureaucracy was a creature born of the fossil-fuel age. It came to maturity in the 1920s, with the shift from coal to oil and from steam-powered to electrified factories. While bureaucracies of various kinds existed in previous civilizations, the new business form of bureaucracy was in many ways unique. The great 20th-century sociologist Max Weber set out to define its features. Among its essential characteristics were predetermined rules for governing decision making, authority exercised from the top of the structure down, well-defined and -written job descriptions for every level of the organization, objective criteria for evaluating performance and advancement, and a division of labor into specialized tasks and functions. This kind of rationalized governing process, noted Weber, made it possible to oversee large, complex organizations, integrating multiple activities under a single roof at ever-faster throughput and speeds.

Many other rationalized mechanisms evolved in the transition years, leading to a fully mature industrial capitalism. Standardized time zones, for example, were introduced, first by the railroads, to better regulate the flow of traffic. In 1870, a rail passenger traveling from Washington to San Francisco would have to reset his watch more than 200 times to stay current with all the local time systems across the country.[65] The great divergency in local time zones created havoc for the railroads in scheduling passenger trains and routing freight. In 1884, standard time zones were established for the world, with Greenwich, England, selected as zero longitude.[66]

Other rationalized processes were hurried into place to support the new bureaucratic structures and help expedite the accelerated throughput of economic activity, including the standardized grading of commodities, standardized machine packaging, and standardized pricing of products at the retail level. Continuous flow production—the first automated factories—churned out cigarettes, matches, soup, and flour in exact units and unvarying composition. New marketing tools, including mail-order catalogs and product branding, as well as new retail mechanisms like franchises—which were first introduced by International Harvester and the Singer Sewing Machine Company and later by the auto industry—turned commercial life into an accelerated flow of standardized goods that were predictable and unwavering in quality.

Frederick Winslow Taylor, the first management expert of the new century, introduced his principles of scientific management into American factories and offices. His theories were later extended into the home, the school, and virtually every other area of contemporary life. Taylor set out to rationalize human behavior itself, to mold it to the rationalized new bureaucratic forms of organization that were remaking American commercial life.

Taylor's goal was to make each worker more efficient by applying principles used by engineers in the improvement of machines. Using a stopwatch, Taylor divided each worker's tasks into the smallest identifiable operational components, then timed each to ascertain the best time attainable under optimal performance conditions. His time studies calculated worker performance to fractions of a second. By studying the mean times and best times achieved in each component of the workers' jobs, Taylor could make recommendations for how best to change the most minute aspects of worker performance in order to save precious seconds.

Taylor believed that the best way to optimize the efficiency of each worker was to assert complete control over all six temporal dimensions: sequence, duration, schedule, rhythm, synchronization, and time perspective. No aspect of the workers' time was to be left to chance or worker discretion.

Instead, the workers were to be placed under the total control of management. They were to be stripped of the knowledge and skills necessary to oversee their work and made into automatons, like the machines they attended. "If the workers' execution is guided by their own conception," argued Taylor, "it is not possible . . . to enforce upon them either the methodological efficiency or the working pace desired by capital."[67] Management's tool for exercising total control over the worker was the work schedule. Taylor wrote:

> The work of every man is fully planned out by the management at least one day in advance, and each man receives in most cases complete written instructions, describing in detail the task which he is to accomplish, as well as the means to be used in doing the work . . . this task specifies not only what is to be done, but how it is to be done and the exact time allowed for doing it.[68]

Taylor popularized the idea of human efficiency and turned an entire nation into efficient human machines. He succeeded in quickening the pace of human activity to match the pace set by coal- and oil-powered machinery and electrified technologies in factories, offices, and retail commercial enterprises. Henceforth, maximizing output in the minimum time with the minimum input of labor and capital became the watchword of commercial life, and soon of private life as well. Perhaps no other individual has had so great an impact on rationalizing human behavior to the dictates of the new fast-throughput high-energy culture as did Taylor. Sociologist Daniel Bell says of Taylor, "If any social upheaval can be attributed to one man, the logic of efficiency as a mode of life is due to Taylor."[69]

The organizational form first developed by the railroads and institutionalized across American industry by Frederick Taylor and a new genre of management experts remained little changed for the next seventy years. Centralized bureaucratic management of integrated business activity became synonymous with American commerce because the institutional arrangement was the best fit for managing a greatly sped-up economic throughput made possible by the conversion to coal and then, later, to oil and natural gas.

Commercial activity in the U.S. and around the world continued to become more and more hierarchical and centralized in direct proportion to the dramatic increase in the energy flow and economic throughput in each successive decade of the 20th century. By the 1960s, the 200 largest manufacturing companies owned 56.3 percent of the total industrial assets of the country.[70]

The birth of the information society in the 1980s changed the way companies did work. The more traditional hierarchical organizational model was found to be too slow to accommodate the quickened pace and density of communication and commercial activity made possible by the computer and telecommunications revolutions. Personal computers, palm pilots, cell phones, and the World Wide Web created instantaneous mobile connections between people. Moving information the old-fashioned way up through level after level of management until it reached the top and then waiting for decisions to move slowly through each level back down to the bottom was becoming increasingly costly in a world where information was moving in every direction at the speed of light.

Companies began to flatten their organizational hierarchies, creating horizontal organizational models that could better accommodate the mobility, flexibility, and speed of a new commercial era. Layers of middle management were reduced or eliminated altogether. The new organizational chart decentralized decision making, delegating more authority directly to personnel at the point of engagement with suppliers and clients.

While hierarchies gave way to networks and a great deal of the operational decision making was dispersed to reduce transaction costs and improve profit margins, the companies themselves became even bigger and more encompassing in their commercial reach in order to manage an increasingly interconnected web of business relationships and activities. Mergers and acquisitions became the norms in every industry. In banking, insurance, telecommunications, utilities, entertainment, pharmaceuticals, agriculture, automotive, steel, and countless other fields (including energy), fewer corporate players concentrated and centralized greater power over their respective industries.

Worldwide mergers and acquisitions have reached new heights in recent years. In 1999, the value of all global mergers and acquisitions was $3.4 trillion, a record and a 40 percent increase over the previous year of $2.5 trillion. Over the past twenty years, the annual value of mergers has risen a hundredfold, reaching a staggering cumulative $15 trillion. In 1999, more than 32,000 separate mergers and acquisitions were announced, triple the number ten years earlier. Today, cross-border mergers represent 33 percent of the total, up 13 percent from the early 1980s and a clear sign of the accelerated process of globalization of commerce.[71]

The sheer size of today's mergers is mind-boggling. The merger of America Online and Time Warner in 2000 was valued at $165 billion.[72] Three industries—telecommunications, commercial banking, and radio and television broadcasting—accounted for one-third of the value of all worldwide mergers in 1999.[73]

Corporate concentration of power over worldwide commerce and trade continues to increase. With each passing year, fewer players dominate the international economy. Of the 100 largest economies in the world today, fifty-one are corporations, while only forty-nine are countries (based on a comparison of corporate sales and countries' gross domestic product). The 200 largest companies' combined sales are bigger than the combined economies of every nation in the world, excluding the top ten nations. The 1999 sales of *each* of the top five corporations exceeded the GDP of 182 nations. Despite the popular myth that the biggest corporate players continue to change as old companies fall by the wayside and new upstarts ascend to prominence, the fact is that more than half of the top 200 companies listed in 1983 were still on the list in 1999, although their names may have been altered or changed because of mergers.[74]

By the opening decade of the 21st century, many critical industries were controlled by fewer than ten global companies. The consolidation and centralization of economic power have continued to move in tandem with the increase in energy flow, and they will likely to so until the global production of oil peaks sometime in the coming decade.

The oil age has been characterized, from the very beginning, by economies of scale. Big oil and big business have gone together. Now, control over global energy and all of the commercial activities that flow from it is clustered around 500 transnational enterprises, many of which are connected upstream and downstream with one another in a dense network of interdependent relationships. Globalization is no longer a future goal, but rather an emerging reality. At no time in history have so many human beings around the world been so dependent on so few institutions for their survival and well-being. Those institutions, in turn, exist by the good grace of the continuous flow of fossil fuels—especially oil—that moves every aspect of modern commercial life.

If, as Leslie White and other anthropologists and historians suggest, the measure of a civilization is, at least in part, a measure of the amount of energy that flows through it, then the fossil-fuel civilization is certainly a high-water mark, as it has consumed the most energy and produced the highest standard of living in history for those who have been its beneficiaries. But, if greater energy throughput has also been synonymous with greater concentration of institutional power over the energy flow and the commercial activity that accompanies it, then it should come as no surprise that the fossil-fuel era has also created the most centralized and hierarchical command-and-control institutions ever to administer an energy regime. The 20th century has truly been characterized by a new kind of empire building, this one founded on oil and managed by giant corporate enterprises working with—and sometimes at odds with—national governments.

The problem is that more concentrated and centralized power in fewer institutional hands has always meant less flexibility in addressing new challenges and more vulnerability to both external and internal disruption. Greater energy throughput has also always resulted in greater entropy in the overall environment. The future, then, of a mature oil era is likely to be at least as costly as its past has proven to be beneficial. As we ascended the bell curve of global oil production, our focus was understandably on maximizing the gains. Now, as we come closer to the summit and begin the long journey down the backside of the bell curve, we need to place as much attention on minimizing losses and preparing ourselves for a new energy regime. The first issue at hand is to understand the enormity of the challenge that lies ahead.

The looming peak in global production of conventional crude oil is being played out against the backdrop of two other potentially destabilizing forces: the rise of Islamic fundamentalism in the Middle East and around the world and the increased warming of the Earth's climate from the burning of fossil fuels. The synergistic effects of each of these three phenomena on the others will be critical in determining the prospects for human civilization in the coming century.

5

THE ISLAMIST WILD CARD

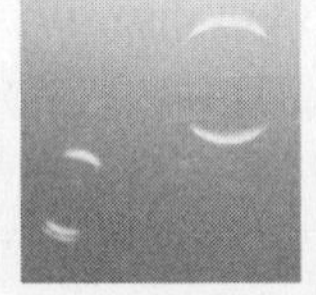

"A soft loan from Allah." That is how many in the Muslim world have come to regard the vast oil riches that lay buried just beneath their feet. Radical Islamists like to point out that, wherever Muslims congregate in large numbers, the Earth they walk on is gushing at the seams with black gold. It's as if Allah has anointed the very ground his earthly messengers trod, making it a hallowed place, replete with religious and historic significance. Serendipitous or ordained, the congruence is rather remarkable. Ten of the thirteen OPEC nations are Muslim states: Saudi Arabia, the United Arab Emirates, Qatar, Iran, Iraq, Kuwait, Algeria, Libya, Indonesia, and Nigeria (with a half-Muslim population). Other major oil producers, including Oman, Bahrain, Syria, Egypt, Brunei, Tunisia, and Malaysia, are also Muslim countries.[1]

At the center of oil and Islam is Saudi Arabia, which holds the largest oil reserves in the world and is the holy land of Islam—the birthplace of the prophet Mohammed and the gatekeeper of the sacred Muslim shrines of Mecca and Medina. While hardened geopolitical analysts in the West snicker at the idea that Allah bestowed such a gift on the upholders of the faith, no one dares laugh when Osama bin Laden urges his followers around the world to reclaim the Saudi holy land, establish a universal Islamic state, and raise the price of oil to $144 a barrel.[2]

The story of oil gives credence to the idea that "what goes around eventually comes around" in history. Oil, the energy that helped make the West the unchallenged economic, political, and cultural force in the world in the 20th century, could now become its undoing at the hands of an Islamic world determined to turn the tables and restore its former status as the world's spiritual and cultural arbiter. Of this much we can be sure: Oil and Islam are inseparably linked. The fate of one will, to a great extent, determine the fate of the other in the coming century.

There are currently 1.2 billion Muslims in the world, comprising 20 percent of the human race. They are a majority in fifty-one countries and a sizeable minority in many others. Muslims are the fastest-growing population group in the world. Demographers forecast that, by 2025, one out of every four people on the planet will be a Muslim. If demographics is power, then the world is tilting toward a Muslim century.[3]

Although their numbers are increasing, their economic fortunes have waned. According to the World Bank, the average income in what is known as the Islamic Belt, stretching from Morocco to Bangladesh, is less than $3,700 a year, half the world average of $7,350.[4] Even worse, the countries of the Islamic Belt are not just poor but are continuing to lose ground to other developing nations. For example, in 1950, Egypt and South Korea enjoyed approximately the same standard of living. Today, South Korea's is five times that of Egypt's.[5]

Oil is increasingly viewed by a younger generation of Muslims as the "great equalizer," a spiritual as well as geopolitical weapon that, if Islamized in the service of Allah, could lead to the second coming of Islam. King Fahd of Saudi Arabia sensed as much in the aftermath of the oil shock of the 1970s and early 1980s, and he told his fellow Muslims that "the main resource to depend on after God is oil."[6]

The great resurgence of Islam among the youth, and the accompanying Islamization and politicization of oil, is the latest chapter in a 1,500-year history of confrontation with Christianity and the West, in which the Islamic world has been, at times, both victor and vanquished, ruler and ruled. For many Muslims, who have experienced only a sense of defeat and humiliation at the hands of the Western powers for most of the 20th century, the prospect of controlling the pipeline to the last remaining crude-oil reserves in the world is a chance to square accounts. For the Western powers, the energy companies, the global business community, and the consumers, the very thought that Saudi Arabia and other oil-producing countries in the Gulf might fall into the hands of a younger generation of fundamentalists seeking to impose a universal Islamic state on the world—a Muslim slant on globalization—is frightening.

Mohammed's Vision

The current fundamentalist resurgence is part of a complex historical narrative of which many of us in the West are ignorant. What needs to be appreciated is why so many young Muslims in the Middle East and elsewhere, both the poor and the well-off, the uneducated and the schooled, are finding common ground in what they call a "spiritual rebirth" and what we tend to regard as a menacing political polarization of the world. Understanding the history of Islam, from a Muslim perspective, provides clues as to what the future is likely to bring when both global oil production and the youth rebellion within Islam peak in the next decade.

The first thing to know about Islam is that while, like Christianity, it traces its roots to Judaism and is in the great tradition of monotheism that began with the prophet Abraham, it diverges in one fundamental respect: its understanding of the role of the pious in earthly history. For Christians, the earthly sojourn is of far less consequence than eternal salvation in the world to come. Christianity, from the very beginning, has been an otherworldly religion. Early on, St. Augustine and other church leaders made it clear that this world is of lesser consequence and that the time spent here is to be used to spread the good news of the coming of Christ and prepare for the world to come. While Christians were to serve witness to the coming of God's kingdom by being his stewards here on Earth, daily affairs of history were to be left to the "powers and principalities." "Give unto Caesar that which is Caesar's" became the great watchword of the faith. If human beings and the rest of the creation were truly fallen—immersed in original sin by dint of the fall from grace in the Garden of Eden—then restoration would only be found in the world to come.

Islam, however, was born of a different inspiration. Its founder, the prophet Mohammed, looked for the presence of God in human history and believed that history itself was the critical arena in which human beings played out their relationship to God. This was a far different notion than the one that inspired early Christian believers to separate from worldly affairs and create the monastic orders of the medieval era.

The prophet Mohammed was born in A.D. 570 and grew up a member of the Quraysh tribe that flourished at that time in the city of Mecca. In midlife, he became anguished over the effects that newfound wealth was having on his tribe, especially the leaders. Many tribesmen were becoming obsessed with making money, often at the expense of their neighbors. They were also less charitable and unwilling to share their good fortunes with the weaker members of their tribe, creating new divisions that threatened to undermine and destroy the very social fabric that had long kept the tribe together. God heard his anguish and came to him on the night of 17th Ramadan in A.D. 610. Mohammed awoke from his sleep feeling a deep presence engulfing him. The first words of the new Arab scriptures came pouring from his mouth. In the years to come, Mohammed experienced God's presence again and again, and each time came new Qur'anic verses. He began to share the words, first with close relatives, then among friends, and eventually with strangers from near and far.

The Qur'an (recitation) was, above all else, a guide to practical living in the world. It called upon the faithful to create a just and humane society, one that ministered to the weak and took care of the poor, and in which all people lived with love and respect for their fellow human beings. While Judaism, Christianity, and the other great monotheistic religions also preached economic and social justice and extolled the virtues of practicing the Golden Rule, what made Islam unique was the emphasis it placed on living one's beliefs in the rough-and-tumble world of day-to-day politics in the social arena. The mission of every upholder of the faith was to redeem history by creating a society that mirrored one's faith. In this new scheme of things, the spiritual and temporal kingdoms were to be experienced as one and the same. To live a spiritual life is to lead a just life in a compassionate society. Conversely, by living justly and with compassion in society, one becomes spiritual.

Mohammed's message fit the tenor of his times. In the world of Mecca and Medina, where newfound wealth was making some rich while impoverishing others, and where traditional obligations and responsibilities to kin and neighbors were being threatened by greed and material gain, the poetic verses of the Qur'an, with their eloquent appeal to live a more just life, fell on welcome ears among both the alienated well-off members of society and the weak and poor.

Mohammed envisioned a universal *ummah* (community) of the faithful, united by a collective commitment to create a just and equitable society. Toward that end, the Qur'an laid out very specific instructions for how the followers were to behave. For example, every Muslim is required to donate a portion of his income to the poor. During the holy days of Ramadan, the faithful fast to feel the pain and suffering of those who do not have sufficient food and drink to fill their own stomachs.

If the ummah was successful, if it grew and prospered, it must be a sign that the faithful were living life in accord with God's will. On the other hand, if the ummah was rife with animosities and divisions and plagued by suffering, it could only mean that the people were not obeying God's will.

Within less than a century after Mohammed's death, Islam had expanded its reach from the Pyrenees to the Himalayas, uniting diverse peoples into a new universal brotherhood and creating a vast new empire whose influence would still be felt 1,300 years later. The conquests reinforced the idea that God's will was being fulfilled. Military and political success allowed every Muslim to feel a part of a transcendent experience.

Islam, then, is fundamentally different from Christianity. It does not separate the city of man from the city of God. The true Muslim lives in a seamless world. His day-to-day life is devoted to building the ummah—a universal brotherhood—that will reflect God's will and, by so doing, be a sign of transcendence.

By dividing earthly and eternal existence into separate realms, the Christian world was able to create the conditions both for the emergence of the independent secular state and the privatization of faith. Islam makes no such separation. Its politics and theology are intimately connected. To lead the correct political life is to lead the correct spiritual life. In this sense, Islam is a truly universal vision. But, because life is, at the same time, both political and spiritual, and thus indivisible, there can never really be a separation of church and state, or reason and faith—the kind of bifurcation that helped create the modern Western frame of mind.

For much of its early history, Islam prospered under this unified vision. At its height, the Islamic empire rivaled that of ancient Rome. While western Europe descended deeper into the Dark Ages and by the eighth century was nearly moribund, Islam was becoming the undisputed seat of learning and cultural advance. Nowhere was Islam's prominence more evident than in the fields of philosophy and science. Muslims rediscovered ancient Greek philosophy, mathematics, and astronomy and created the early prototypes of what would later become the modern university. For the next 500 years, Muslim scholars were involved in scientific pursuits. Their libraries in Cordoba and Toledo, Spain, were the envy of the world.

The question, then, is why was it the West, and not Islam, that eventually succeeded in harnessing science to the task of conquering nature and colonizing much of the world in the modern era?

The answer lies in the indivisible nature of the Islamic vision. For Muslims, good science, like good politics, ought to reflect the glory of God's presence and help the faithful lead a more just and spiritual life. Greek philosophy, to the extent that it continually inquired into the why of things in order to better understand the world as it has been created, made sense. Farouk el-Baz, a geologist at Boston University and formerly science advisor to the late President Anwar Sadat of Egypt, explains the Muslim approach to science: "When you know more, you see more evidence of God."[7] Muslims look to science as a way to better understand the unity of existence that is God's handiwork. That is not to say that science should not also be used for practical purposes. For example, the great interest in astronomy was sparked, in part, by the need to make sure that each follower would know which direction to face each day to be in line with Mecca during prayer. Muslim scholars produced elaborate diagrams and tables by which sacred directions could be ascertained for Muslims everywhere in the world.[8] Although utilitarian at times, science was, and is still today, seen as a window to the divine. For Muslim scholars, science was never meant to be put into the service of men, but rather to serve God.

The European Approach to Science

The West, by contrast, underwent a long and protracted struggle that eventually separated science from any otherworldly considerations and, in so doing, unleashed a Promethean spirit that would remake the modern world in the image of Western materialism. Max Weber calls this epic struggle "the disenchantment of the world." Again, the roots of this disenchantment are found early in the Christian experience and lie in the clear distinction made between two realms of existence, the earthly and the eternal. Christians experienced the world very differently than did Muslims. While the latter lived out their faith in the temporal world, Christians saw themselves as being in this world but not of it. Earthly existence was to be tolerated but not to be reveled in. It had no transcendent aspect to it. Remember also that the first Christians expected Christ's imminent return and were therefore not much concerned about their earthly sojourn. When the Second Coming did not occur, the Church elders began to institutionalize their hold over earthly affairs by forming alliances with the local "powers and principalities" to ensure a steady stream of revenue into the Church coffers. But the presence of the practical arts, which were mostly attended to by monastic orders, were not considered to be of much significance and were therefore largely left alone.

By the 13th century, however, Christian Europe was slowly reawakening from its long sleep. New agricultural methods were increasing food production, providing a surplus for the first time since the fall of the Roman Empire. A fledging commerce flourished, trade routes began to open up, villages gave way to towns and small cities, and craft production expanded into new industries. All of this development forced a crisis of sorts for the Church, which had traditionally tolerated the practical arts as long as they were relegated to a secondary status. The new earthly material interests, however, were becoming a challenge to the faith. The great church scholar St. Thomas Aquinas attempted to address the potential schism between the temporal and spiritual realms in his *Summa Theologica* in the 13th century. St. Thomas opined that man was granted two higher faculties by God: reason and faith. Reason allowed man to explore the material world, in order to better understand God's creation, and therefore ought to be encouraged. However, it should be understood to be of a lesser nature than faith. What reason cannot answer is left to the province of faith to address. St. Thomas's compromise, which scholars called "The Delicate Synthesis," was designed to allow reason to flourish in the practical realm as long as it did not attempt to usurp faith in the spiritual realm. If reason were to prevail over faith, then man would be asserting his sovereignty over God's. St. Thomas's attempt to hold the line did not succeed.

The Protestant Reformation in the 16th century, while intended to purify the faith against earthly corruptions, had the unintended consequence of furthering the schism between the material and spiritual realms. Martin Luther and John Calvin preached a new catechism of earthly callings. They both argued that the Church's teachings that stressed "good works" could not ensure any believer a place in heaven. According to the reformers, every individual is either elected to eternal salvation or dammed to hell at birth. Still, said Calvin, one has a responsibility to act as if one has been selected by pursuing a "calling" and working tirelessly to improve one's lot, as a way to reassure oneself, against one's own gnawing doubts, that one is among the blessed. Henceforth, the faithful organized their lives in the most methodical way possible to achieve success, not for love of personal gain but as a "sign" of their election. The new emphasis on achievements in this world gave impetus to a more materialist ethos. Max Weber referred to the new way of living as the "Protestant ethic" and argued that it inadvertently produced a rational frame of mind tailor-made for industrial pursuits and modern capitalist behavior.

By the 1600s, the two realms—the earthly and the eternal—were at war on a number of fronts. Reason was being harnessed by a new generation of philosophers committed to challenging many of the most treasured doctrines of the Church. Francis Bacon wrote his *Novum Organum* in 1620, extolling the virtues of what he called "the scientific method," a way to pursue knowledge that would unlock the secrets of nature and make man the master of the universe. Bacon was particularly hard on the Greek science that Muslim scholars so revered. The Socratic tradition, with its emphasis on the why of things, was, to Bacon's way of thinking, utterly unsuccessful in enhancing the material well-being of humanity. The Greeks, he ranted, had not "adduced a single experiment which tends to relieve and benefit the condition of man."[9] The ancient Greeks, Bacon wrote derisively, "assuredly have that which is characteristic of boys; they are prompt to prattle but cannot generate; for their wisdom abounds in words but is barren of works."[10] Bacon was far more interested in the practical benefits that science could bestow. He viewed nature not as a window to the sacred but rather as a common "harlot" and urged future generations to "shake her to her foundations" to "squeeze," "mould," and "shape" her in order to "enlarge the bounds of human empire to the effecting of all things possible."[11] Bacon's ideas about nature and the role of science would be an anathema to Muslim scholars. His was a purely utilitarian philosophy designed to assert power over the creation and optimize material gains for society.

Seventeenth-century Europe, however, was more than ready to hear such heresy. The newly emerging mercantilist economies, with their far-flung colonial adventures, were themselves busy taming the natural world and enlarging the bounds of human empire. Bacon's ideas about nature and science were greeted with enthusiasm.

The Church, on the other hand, was ambivalent. Although anxious to Christianize the savages in the New World, the Church leaders were uncomfortable with the humanist bravado that accompanied the colonial quest. The Vatican fought to contain the new rational materialism that was creeping into European thought and influencing ideas about science and nature, but the effort was only half-hearted. The Church attempted to silence the Italian astronomer Galileo from advancing his argument that the Earth revolves around the sun and sentenced him to life imprisonment for heresy, only to quickly commute his sentence to permanent house arrest. By then, Western material progress was too far along to be slowed down by Church doctrine. Faith gave way to reason, and within a century philosophers were proclaiming the coming of a new material kingdom here on Earth. The Marquis de Condorcet, the French philosopher, remarked at the onset of the French revolution that

> no bounds have been fixed to the improvement of the human faculties . . . the perfectibility of man is infinite . . . the progress of the perfectibility henceforth above the control of every power that would impede it, has no other limit than the duration of the globe upon which nature has placed us.[12]

The West had successfully separated reason and faith and had made the first dominant in the affairs of society while relegating the second to private life.

The Church could continue to advance the faith but would no longer be the overarching organizing schema for society. In its place would be a new institutional model, the nation-state, a secular vehicle tailored to spread "reason"—in the form of modern science, technology, and commerce—to the four corners of the Earth. The new trinity would come to rule the world over the course of the next two centuries, and, everywhere its tentacles reached, Islam would face retreat, subjugation, and, ultimately, humiliation at the hands of the West.

By the late 19th century, many Muslims were beginning to ask what the West possessed in the way of educational models, technologies, and political and secular institutions that allowed it to exercise such a commanding influence in the world. The final breakup of the Ottoman Empire after World War I convinced many Muslims that the only way to slow the retreat and regain a global foothold for Islam was to learn how to expropriate the things that the West had that made it powerful and adapt them to their own circumstances.

Western Influences

Two groups emerged early in the 20th century: the Westernizers and the Modernists. The first attempted to borrow, carte blanche, the Western model, and place it in the Islamic world—a structural graft that would challenge the underlying orthodoxy of Islam at its core. Muslim scholars and theologians, especially in Egypt and India, created a kind of Islamic version of the European Enlightenment. They attempted to revise Islam in the Western mold by urging deep structural and social reforms, including the separation of religion and state, the establishment of a civil society, the separation of science from religious oversight, the creation of an independent judiciary free of the long arm of religious authority, and the establishment of electoral politics. The Modernists were more circumspect. Their hope was to borrow selectively from among Western political, educational, and legal models and to try to adapt them to Islamic thinking without losing the essence of Islam. The Westernizers won out, in no small measure because Western colonial influence was already so pervasive in the Muslim world. Between the end of World War I and the 1960s, Muslim elites in the Middle East and elsewhere were educated in Europe and America and adopted Western dress and lifestyles. The Middle East was Europeanized and then Americanized. The nation-state model was introduced across the region, and a new generation of political leaders began the process of creating a secular culture.

The history of the Westernization of the Middle East has been one of continuous failure. The model was simply too alien to the Islamic experience of ummah to be successful. Getting Muslims to accept the nation-state—a secular model of governance based on arbitrary assignment of political boundaries—when for centuries Muslims' deepest loyalty has always been to a universal brotherhood that is boundaryless and based on religious faith—was an impossible task. That is not to say that Muslims weren't still tribal and territorial; however, their larger affiliation has always been to the ummah. The nation-state was and is still seen by many in the Middle East and elsewhere as a colonial institution designed to divide and conquer Muslims.

Their perception is borne out in the history of the region. European powers, especially the British, maintained a colonial presence in the Middle East for more than 100 years. During the colonial era, nation-states were established, often with the help of local strongmen, reflecting the interests of European powers and later the United States. These so-called nation-states were designed more in an effort to consolidate trade routes, commandeer vital resources, and protect strategic military interests than to align like-minded peoples in common political pursuit. Often, political boundaries were superimposed by force on diverse local tribes, generating dissension and animosity that then had to be periodically quelled by police and military intervention.

For a time, after World War II, a homegrown nationalism, in the form of pan-Arabism, flourished. Inspired by the anticolonial struggles in India, Africa, and Asia in the late 1940s and early 1950s, a new generation of Arab leaders, led by Gamal Abdel Nasser of Egypt, began to call for an alternative approach to nationhood. Distancing himself from the colonial form of nationalism that had been imposed on Egypt in the first half of the century, Nasser consolidated power in the 1950s and introduced a secular regime that was socialist in bent and modern in form. Nasser dreamed of uniting the Arab world into a single Arab-speaking nation, much the same way nationalists had united the German- and Italian-speaking regions into single nations eighty years earlier.

Nasser's pan-Arab vision swept the Middle East. After a century of colonial usurpation, a new generation of leaders was in desperate search of an alternative vision of its own. The initial enthusiasm did not last long, however. Pan-Arab nationalism soon became tainted by a new kind of colonial dependency, this time to the geopolitical agenda of the former Soviet Union. Substituting Western colonialism for Soviet colonialism seemed a paltry tradeoff. Then, too, Egypt's socialism was unable to develop either the country or the region. By the mid-1960s, many of Nasser's commercial plans had failed. Egypt and other countries found themselves sliding inexorably backward. The grand vision of a pan-Arab approach to modernization and nationhood faded. The coup de grâce for pan-Arabism came in the form of a humiliating defeat by Israel in the 1967 war. If the whole Arab world was unable to defeat a small nation of only three million Jews, then talk of a pan-Arab resurgence seemed foolish.

Nasser's brand of pan-Arabism, laced with socialism, had a profound effect "on the street." Nasser realized that his dream of a pan-Arab state depended on his ability to loosen the hold of Islam on the day-to-day life of the people. Secularizing Egyptian society inevitably brought Nasser and his government into direct confrontation with upholders of the Islamic faith. The Muslim Brotherhood, the country's most visible Islamic group, was singled out by Nasser for particularly harsh treatment. More than 1,000 of the organization's members were imprisoned in concentration camps, many for years. Other Middle Eastern regimes, under the banner of modernization and secularization, were equally severe in their treatment of Islamic groups. In Iran, Muhammad Reza Shah closed *madrasahs* (Islamic schools), tortured, imprisoned, and exiled *ulama* (religious leaders), shot Islamic dissidents, and carried out a systematic public campaign aimed at purging Iran of Islamic influences, from dress codes to lifestyles. These and similar efforts in other Middle Eastern countries served to further embitter Muslims about the idea of a secular society.

Modernization, at the hands of the new nation-state, had yet another profound, far-reaching effect on the population—especially the young. In Egypt, Iran, Jordan, and elsewhere, the rush to modernize the economies led to a depopulation of rural areas and a migration into cities. Dislodged from their native soil and local communities and forced to relocate in densely overpopulated urban areas like Cairo and Tehran, many Muslims lost a sense of mooring. The secular society of the city provided little or no solace. Impoverished, isolated, and stripped of a sense of personal identity, many young Muslims came to think of themselves as aliens living in a hostile world. Pan-Arabism, socialism, secularism, and modernism had all failed them. In search of an answer to the misfortunes being heaped on them, a growing number of Muslims turned back to their faith for guidance. Many found what they were looking for in the form of a new movement, Islamization.

Islamization

The spiritual father of modern Islamic fundamentalism is an Egyptian, Sayyid Qutb. As a young man active in the Muslim Brotherhood, Qutb hoped to find a way to adopt Western democracy to the Islamic context without losing the spiritual essence of the faith to secular influences. After being imprisoned by Nasser for his activities in the Brotherhood, Qutb had a change of heart. Witnessing the brutal torture and murder of his "brothers" at the hands of Egyptian police, Qutb came to believe that Islam was incompatible with the tenets of a secular society.

While in prison, Qutb wrote a book, *Signposts*, in which he condemned Nasser and the leaders of virtually every other Arab country at the time for being *jahiliyya*. Originally the term had been used to refer to pre-Islamic Arabia, the period Muslims regarded as "The Age of Ignorance." Jahiliyya meant "the enemies of faith," the forces of barbarism who refused to submit to God's will. Accusing Arab leaders of jahiliyya was a political bombshell.

Qutb was saying, in effect, that the leadership of the Arab world was infected by the poisonous influences of the West and were actively pursuing policies designed to turn the faithful away from Islam. Further, Qutb argued that true Muslims had an obligation to overthrow such regimes. His words amounted to a declaration of rebellion against those in power inside the Muslim world. Its message slowly began to reverberate across the Middle East. Islamic fundamentalism had been born.

Although there are different strains of contemporary Islamic fundamentalism, there are a few broad themes to which every fundamentalist sect cleaves. First, the reason the Muslim world is in decline is that the people and their rulers have abandoned the teachings of Mohammed and the Qur'an. Second, the weakening of the faith has allowed pernicious influences from the West to penetrate Muslim life and take hold throughout the Muslim world—especially materialism, secularism, nationalism, and immoral ways of living. Third, the solution to the problem is the re-Islamization of the Muslim world. This means, among other things, the re-establishment of *shariah* (Muslim law), a strict adherence to traditional Islamic codes of behavior and the purging from society of Western influences, especially its decadent lifestyles and cultural values. Western technologies and some Western commercial forms, however, can be retained if Islamized. Fourth, re-Islamization of society can only be accomplished by repoliticizing Islam. Islamic fundamentalists are often most critical of the ulama itself—the guardians of the legal and religious traditions of Islam—arguing that they have lost the sense of political engagement, which is so essential a part of Islam's mission. Many religious leaders, they claim, have become apolitical or, even worse, apologists for corrupt regimes and not fit, therefore, to lead the struggle to create a just society. Thus, many of the fundamentalist movements are lead by laymen, not religious leaders.[13]

Young Muslim fundamentalists today chant "Islam is the solution." The first inkling that Islamization might pave the way to a new beginning for the faithful came in the Arab-Israeli war of 1973. Seven years earlier, Muslims had gone to war with the battle cry "land, sea, sky," pinning their hopes on sophisticated military equipment. They were crushed. In 1973, the battle cry shifted from technology to faith. "God is great," chanted the warriors.[14] Their renewed faith appeared to pay off with success on the battlefield. At least, that is how many young fundamentalists came to see the event.

The Iranian revolution in 1979 is viewed by many as the seminal event that ignited the current Islamic revival and gave impetus to fundamentalist movements throughout the Middle East and around the world. The shah of Iran, for many Muslims, represented the worst excesses of Western influence in the Middle East. His overthrow and replacement by a fundamentalist regime, under the tutelage of the Ayatollah Khomeini, provided the fundamentalists with the first concrete example of a political success. The dream and the reality had come together, making the re-Islamization of the Middle East a real possibility for the first time.

Another force, however, perhaps far more significant, was emerging in the 1970s that would convince millions of Muslims in the Middle East and elsewhere that re-Islamization was destined to change the course of history. The great oil boom of the 1970s profoundly changed Muslims' sense of themselves as well as the West's image of the Muslim world. These changes in perception have played a critical role in redefining Islam, both internally and in its relationship to the rest of the global community.

To grasp the enormity of the change, we have to remember that in 1970, crude oil was selling at a relatively stable price of $3 per barrel on world markets. Ten years later, a barrel of oil was selling for $34.[15] The increase in the price of oil was matched by equally dramatic increases in production. Saudi Arabia was producing 3.5 million barrels of oil per day in 1970. By 1980, it was selling 10 million barrels a day.[16] The new wealth flowing into the coffers of oil-producing countries was without precedent. Saudi Arabia took in $1.2 billion in oil revenue in 1970. A decade later, it was taking in more than $101 billion per year.[17]

The shift from a buyer's to a seller's market caught the rest of the world unprepared. All of a sudden, there was a dawning awareness that oil, the lifeline for the entire global economy, was in the hands of people and governments that previously had been regarded as little more than commodity-providers. Arab sheiks were courted by world leaders anxious to curry favor. Global companies were flying their senior executives into the Middle East on private jets, lobbying governments to secure commercially lucrative business contracts. As Daniel Pipes points out in his book *In the Path of God*, "just a few years earlier, the notion that the Arab nations might rank among the world's wealthiest would have been preposterous; when sudden riches came, they appeared all the more awesome and even miraculous for being so unexpected."[18]

Pipes says that the oil boom had a number of other equally interesting effects that buttressed the new confidence in Islam. Governments, wealthy oil sheiks, and other oil-related business interests, flush with newfound wealth, poured funds into missionary work abroad, including the establishment of madrasahs all over the world. Books and other publications espousing Islam were similarly dispersed. Funds were channeled to religious institutions and organizations both at home and abroad, quickening the pace of re-Islamization.

The Muslim world had long sensed that oil might some day help write a new chapter for Islam. It was not until the fall of 1973, however, that the dream of re-Islamization became real. Within the span of just a few weeks, the Muslim regimes of the Middle East turned upside down the carefully constructed world economic order forged by the U.S. and its allies after World War II, and, in the process, became an equal player, at least for a time, on the world stage.

Egypt and Syria attacked Israel on Yom Kippur, the holiest of Jewish holidays, setting off the fourth Arab-Israeli war. Ten days later, OPEC delegates, meeting in Kuwait City, made the decision to raise the price of oil by 70 percent to $5.11 per barrel. As significant as the price increase was the fact that, for the first time, OPEC delegates made the decision unilaterally and without the consent of the oil companies. For a half century, Middle East producers had fought to free themselves from the yoke of Western governments and oil companies. Daniel Yergen, in his book *The Prize*, reminds us that the struggle for independence had been long and tortuous, from the early period in which oil companies set the price, to the days when the exporting countries could exercise a veto, to an era of mutually negotiated prices. Now OPEC had dared to act alone. Ahmed Zaki Yanami, the Saudi oil minister, took stock of the import of what had taken place. "The moment has come," proclaimed Yanami. "We are masters of our own commodity."[19]

Three days later, President Richard Nixon announced a proposed military-aid package of $2.2 billion to Israel, enraging Middle Eastern leaders. Later that day, Libya announced an embargo on all oil shipments to the United States.[20] Saudi Arabia and other oil producers quickly followed suit.

Besides cutting off oil to the United States, the oil embargo set production restrictions. By December, 4.4 million fewer barrels of oil were being produced per day.[21] With oil supplies already tight in world markets, the dip of nearly 9 percent in production sent oil companies and consumers scrambling for oil.[22] Panic set in as the whole world began to realize exactly how dependent they had become on Middle Eastern crude oil. The price of oil on world markets skyrocketed. In the United States, gasoline prices rose by 40 percent, and long gas lines began to appear for the first time as worried motorists waited to fill up their tanks, not knowing if there would be any gas available a week down the line. In December, Arab oil ministers met again and officially raised the price of crude to $11.65 per barrel—a fourfold increase in the price of oil since the onset of the war two months earlier.[23]

The oil embargo seriously threatened the Western Alliance. European powers, in particular, were anxious not to further upset the oil sheiks and reminded the United States that they were far more dependent on Middle Eastern oil than even America was. In November, the European Community passed a resolution in support of the Arab countries in the Arab-Israeli conflict. The gesture was enough to secure for them a suspension of the oil cuts imposed for December by the Arab countries.[24] Japan, even more dependent than Europe on Middle Eastern oil, passed a similar statement in support of the Arab countries in their war against Israel.[25] Japan was similarly rewarded with a suspension of the December cutback.

The dramatic rise in oil prices sent the industrialized nations into a deep recession. Unemployment doubled, and the GNP dipped for the first time since the end of World War II.[26]

The psychological impact of the oil embargo would come to be as important as its economic and geopolitical impacts. The world of Islam, after centuries of decline, had ascended virtually overnight to a position of world prominence. Udo Steinbach, director of the German Institute for Middle East Studies in Hamburg, summed up the profound implications of the change that had occurred in the psyche of every Muslim Arab. He wrote:

> Herein lies the significance of petroleum: The dependence of a majority of the industrial world on Arab (and especially Saudi)

> oil imbued the Muslim Arab with a degree of power and international respect which he needed in order to appreciate the validity of his (political) religion. . . . In this way, he regained his full identity and he found the world once again properly ordered.[27]

Ironically, the oil boom also had the effect of bringing Western modernity to the Middle East at an accelerated pace. Arab countries held court to Western arms dealers, purchasing the most sophisticated and expensive equipment available. Consumer goods and technologies flooded the Middle East, as did Western dress and lifestyles. Middle Eastern youth—financed by government gifts and grants—migrated in droves to European and American schools to advance their educations. So, while the flow of oil moved from the Middle East to the West, the flow of Western goods and ideas flowed the other way. It all happened so quickly that many Muslims experienced a kind of cultural "future shock" and a subsequent identity crisis. The question of what it meant to be a Muslim, in a world surrounded by Western influences and temptations, led to a search for answers. In their search, many young Muslims turned to the teachings of Mohammed.

Recall, the prophet Mohammed had beseeched the faithful to address inequities by creating a just and compassionate society. All around them, young Muslims saw unimaginable wealth enjoyed by only a few in a world where most Arabs still lived in squalor and despair. The newfound wealth had brought prestige to the Middle East, but the regimes were rife with corruption. Control over oil allowed Muslim political leaders to exert great influence abroad but also to wield a bigger political stick at home. The new commanding influence over oil had liberated the Middle East from Western commercial hegemony but had done little to free millions of Arabs from dictatorial regimes. Many young people found what they were looking for in the Islamic revival and the new fundamentalism. Their quest henceforth would be to re-Islamize the Middle East and the world.

Tough Times

The oil surge of the 1970s gave Arab Muslims a renewed sense of their collective destiny and helped spawn the Islamic revival, but it was the hard realities that emerged after the collapse of world oil in the mid-1980s that deepened, hardened, and shaped the current fundamentalist drive across the Arab world. If the oil boom helped restore a sense of confidence to Islam, the oil bust helped convince a younger generation of Muslims of the need to take on not only the West but also their own corrupt, autocratic, and unresponsive governments at home. The long and short of it is that, for a while, Arab governments could mollify their own subjects and at least hold the fundamentalist reformers at bay by guaranteeing employment and income, providing basic services, and bestowing gifts, favors, grants, and subsidies of various kinds to appease religious groups, workers, and others. When petro-wealth began to slow and government treasuries began to shrink, public handouts of all kinds were curtailed, and the only thing that has grown in recent years is popular anger and resentment toward regimes that seem increasingly remote and unresponsive to the needs of their people.

The oil numbers are disconcerting. OPEC oil revenues reached $340 billion per year after the 1974 oil embargo. With the fall of the shah of Iran in 1979 and the start of the Iran-Iraq war in 1980, oil revenues bounded to $438.8 billion. Just six years later, oil revenues had plummeted to less than $83 billion.[28] OPEC oil revenues have remained low ever since.

In the boom years, many governments in the Middle East—especially in the oil-rich Gulf, where 90 percent of the oil reserves in the Middle East lie—became essentially rentier societies. That is, oil revenue, rather than taxes, made up the government revenues. It is not hard to understand why. At the high point of the oil boom in the early 1980s, oil exports constituted more than 50 percent of the GDP in the Gulf nations.[29] Although it is difficult for us in the West to imagine paying no taxes to government and having all of our basic needs met by independent revenue-generating sources, that is the way it is in countries like Saudi Arabia, Kuwait, Qatar, Abu Dhabi, and Dubai. In the halcyon years—during the oil surge of the 1970s—Gulf governments became the all-encompassing central players in the lives of their subjects. Nowhere was the government's new expanded role more evident than in employment. Even after the world oil crash of the mid-1980s, the Kuwait government continued to employ more than half of the workforce in the public sector. In Oman in the 1990s, more than 60 percent of the workforce was paid by the government to work on what is referred to euphemistically as "community services."[30]

Public employment is only the beginning of what has come to be a "cradle-to-grave" dependency. Many Gulf countries provide free public education through the university level, free health-care services, subsidized housing, grants and low-interest loans for starting a business, and social security for the disabled and aged. Saudi Arabia and Kuwait even provide subsidized food through government-sponsored cooperatives. Gasoline is discounted, and public utilities like water, electricity, and telephone service are either free or subsidized.

Of course, nothing is ever really free. In return for taking care of their subjects, the governments in the Gulf have come to expect complete and unwavering loyalty to the state. Political disagreement, even of the moderate kind, is not tolerated. Independent labor unions, once a political force in many Middle Eastern countries, have been outlawed, at least in the Gulf. Political parties are likewise illegal in a number of countries, including Kuwait, Oman, Qatar, Saudi Arabia, and the United Arab Emirates, and the only legal political party allowed to exist in Iraq is that of Saddam Hussein. The Iraq broadcast media are under state control, and newspapers are subject to censorship. The governments are run by hereditary elites, leaving little room for alternative political points of view to be aired or publicly expressed.[31]

As long as oil revenue was in excess of government expenditures on services, the Gulf monarchies could buy the loyalty and obedience of the overwhelming majority of their subjects. For the past decade or so, however, falling oil revenues have failed to match rising government expenditures. Mounting public debt and a steady curtailment of services have made countries throughout the region more politically unstable than at any other time in their history and far more vulnerable to insurgent Islamic fundamentalist movements.

While the poor economic situation is, in large measure, due to falling oil revenues, it is made even worse by the dramatic rise in the number of births in the Middle East. This part of the world has experienced a "youth explosion." Population growth in the 1990s averaged 2.7 percent.[32] Even more unsettling is the fact that 40 percent of the region's population is under the age of seventeen. Today, unemployment among eighteen- to twenty-five-year-olds is averaging about 20 percent, creating a political time bomb in every country.[33]

Unable to diversify the economy or provide new sources of employment, Gulf countries are faced with two equally unpalatable alternatives: Either keep as many youth as possible employed in public-sector work by going deeper into debt, or constrict government payrolls and turn large numbers of young people out onto the streets, creating a recruiting ground for the growing ranks of the militant Islamic sects.

Either way, in the short run—that is, the next several years—as Russia and other non-OPEC producers flood the global market with cheap crude oil, real per-capita income is likely to continue to drop. In the southern Gulf countries, real per-capita income today is only 40 percent of what it was at the peak of the oil boom twenty years ago, and it is expected to continue to decline, creating an even greater risk of spreading social unrest and political upheaval.[34]

Saudi Arabia

Saudi Arabians like to say, "My father rode a camel, I drive a car, my son rides a jet airplane—his son will ride a camel."[35] While one-quarter of all the remaining oil reserves in the world lie in Saudi Arabia, there is an almost fatalistic sense among many Saudis that they are a country living on "borrowed time." How Saudis use that borrowed time will likely determine the way the world makes the transition out of the oil age.

Saudi Arabia, the home of the holy Islamic shrines of Mecca and Medina and the birthplace of Mohammed, is a study in contrast between the traditional and modern and between the spiritual and secular life. Here is a society that just fifty years ago was largely populated by nomadic tribes living much the way their ancestors did at the time of Mohammed. In 1950, there was not a single paved road in the whole country. The port city of Jiddah was a walled city of some 50,000 inhabitants. There was no rail link to the capital, Riyadh. There was only one radio station broadcasting when President John F. Kennedy was assassinated in 1963. Today, there are thousands of miles of highways crisscrossing the country, connecting every village, town, and city. Jiddah is now a world-class city of 1.5 million people. Its boulevards are lined with high-rise apartment buildings and office towers. The Saudi airport at Jiddah takes up forty square miles and is 50 percent larger than Kennedy, La Guardia, O'Hare, and Los Angeles International airports combined. The new airport at Riyadh is twice the size of Jiddah's complex and covers a staggering ninety-four square miles. In 1978, there were few television sets and only 125,000 telephones in the entire country. Today, Saudi Arabia boasts 250 television sets per 1,000 people and a satellite system made up of eleven mobile and three fixed earth stations, providing phone access everywhere in the kingdom. The first electrical generating plant wasn't installed until after World War II. By the beginning of the 21st century, all of Saudi Arabia had been electrified and most of its main buildings air-conditioned.[36]

Yet, in Saudi Arabia, women are not allowed to drive automobiles, movie theaters are banned, there are no elected government bodies, and religious courts are used to adjudicate marriages, divorces, and inheritances.

The Saudi government owes its existence to a partnership, of sorts, that was made between a religious and a tribal ruler more than 250 years ago. In 1745, Mohammed ibn Saud, the ruler of the small oasis town of Dariyah, forged an alliance with an Islamic scholar, Abd al-Wahhab, who proselytized for a return to a strict practice of Islam. The arrangement between the two called for the Saudi ruler to accept the Wahhabi interpretation of Islam as the basis for governance, and, in return, for the religious leader to accept the government as the legitimate ruling body. Over the years, the partnership was kept alive by intermarriage between the two families. The founder of modern Saudi Arabia, King Abd al-Aziz al Sa'ud ar ibn Sa'ud, brought together the various territories of the Arabian peninsula under his rule in the early decades of the 20th century, with the Wahhabi sect again playing a key legitimizing role in the new regime.

In Saudi Arabia, the Wahhabi religious leadership is second only to the royal family in influence. The Wahhabi exercise a measure of direct control over education and social issues and consult with the royal family on affairs of state. The government, for its part, finances the building of mosques, underwrites religious activities, and pays the salaries of religious leaders. The ulama is guaranteed a monopoly over religious practices and a measure of control over the social life of the Saudi people, in return for which the government is guaranteed a compliant religious leadership willing to do its bidding.

Saudi Arabia is therefore a strange hybrid—part autocracy, part theocracy. The royal family has, in the past, viewed the Wahhabi presence as an insurance policy of sorts. The government has prohibited civic organizations from developing in the country and has left it up to the Wahhabi religious leadership to be both monitor and arbiter of the social and cultural life of the Saudi people. Interestingly enough, because other forms of civic expression are forbidden, the mosques have become the only place where people are free to congregate, express opinions, and carry on debate. Not surprisingly, then, much of the pent-up anger and angst, especially among the young, is vented by way of participation in religious forums and gatherings. The result is that local mosques have become deeply politicized in recent years. In the future, say some observers, the local religious institutions are likely to be the breeding ground for a new generation of fundamentalists determined to overthrow the state authority, just as they have served in the past as safety valves to maintain the interests of the Saudi ruling family. Much depends on whether the Saudi government can address the ever-worsening economic situation brought on by years of cheap crude-oil prices on world markets.

Saudi Arabia has changed radically in little more than a single generation. Coping with modernity and the process of globalization in a country wedded to a strict Islamic code of social behavior has proven to be a daunting—some say impossible—task. The country's problems stem, in large part, from a too-fast economic growth followed by an equally powerful economic deceleration. World oil prices were responsible for the rise and fall of the economy, making oil both a gift and a curse at the same time.

In 1970, Saudi Arabia had a population of only 6.2 million people, a GDP of $4 billion, and a per-capita income of only $2,800. Today, the population stands at 22.7 million, and 43 percent of that population is under fourteen years of age. The GDP is $173 billion, and the per-capita income is about $8,500.[37]

The labor force has been transformed, in less than thirty years, from one in which 64 percent of those employed worked in agriculture to one today in which less than 6 percent still work in the fields, while 47 percent work for industry and 47 percent work in services of various kinds.[38] Depopulation of rural areas has been among the quickest in the developing world. By 1995, nearly 80 percent of the people were living in urban areas, and one out of every five was living in a city with a population of more than a million people.[39]

Literacy has risen from 15 to 63 percent of the Saudi population. Ten percent of the Saudi youth, however, drop out of school by fourth grade, and only 5 percent of males and 2 percent of females go on to secondary school.[40] And much of what Saudi youngsters do learn is more in the way of religious instruction, ill preparing them for meaningful employment in a diversified economy.

A poorly educated workforce living inside a one-resource economy is an ill omen for the future. The only way the government has found to absorb the new entrants streaming into the labor force is to put them to work in the public sector. In 1997, over half of the Saudi adult workforce was on the government payroll.[41] Financing all of this, as well as all of the other government services built up during the oil-boom years, costs a lot of money—more than $50 billion a year. Because the oil revenues are not able to keep up with the economic and social needs of a growing population, the Saudi government has had to spend more than it takes in, running a budget deficit every year for the past fifteen years.[42]

Add to these woes that it is going to cost upward of one-third of a trillion dollars to upgrade its oil refineries, pipelines, facilities, transport, and domestic infrastructure over the course of the next two decades simply to maintain its oil industry and economy, and the scale of the problem becomes clear.[43]

Saudi Arabia is in trouble. Its problem is similar to that facing all of the other oil-producing countries in the Gulf: how to politically survive the perilous years immediately ahead until global oil production peaks, making the Middle East once again the supplier of last resort for the global economy.

What has Saudi leaders worried is the millions of young people, poorly educated and without prospect of meaningful employment, who must be taken care of by a government already awash in debt. In 1998 alone, Saudi Arabia's GDP declined by 7 percent while its deficit rose to $13 billion. Meanwhile, its per-capita income continues to decline by about 2.9 percent per year.[44]

Whither Democracy?

There is growing concern throughout the Persian Gulf that "the lid is about to blow." Much of the pressure building up from below is economically motivated, to be sure, but it is also politically directed. These are some of the most repressive and tyrannical governments that exist anywhere. According to the World Audit's index of economic freedom, a majority of Muslim countries are ranked among the most restrictive in the world. Five governments—Saudi Arabia, Iraq, Libya, Sudan, and Somalia—are ranked among the eight most repressive regimes in the world.[45]

Muslim youth, angered over their deteriorating economic circumstances, are in no mood to tolerate the repressive policies of their own governments. The governments, for their part, seem immobilized by the intractable economic problems they face and, at the same time, appear unwilling to create venues to allow the public to express its disenchantment and offer alternative visions for the future.

If the governments of the Middle East don't have the answers, then at least the fundamentalists say they do. Young ears are eager to listen to their message, and an increasing number are becoming converts to the newest "ism," Islamism.

For the young, a figure like Osama bin Laden is particularly appealing. They see him speaking out against their own corrupt rulers and an unjust society. His military triumphs give them hope of an ultimate victory over the forces that have created all the problems in the world. Bin Laden, a product of strict Wahhabi education in Saudi Arabia, is viewed by many young people as a great reformer, someone unafraid to take on both the "hypocrites" at home and the "infidels" abroad on behalf of the "true" faith. The fact that he was born to a wealthy Saudi family and chose to give up his privileged position to fight a *jihad* (holy war) has only enhanced his image among young Muslims in the Middle East and around the world.

By tying the fate of Saudi rulers to the hated machinations of America and the Western powers, bin Laden, in a single brilliant stroke, say some observers, created a new and powerful political dynamic for mobilizing a younger generation to Islamic fundamentalism. The opportunity came when the Saudi government allowed U.S. and other Western troops into the kingdom during the Gulf war. Bin Laden denounced the decision in what amounted to laying down the gauntlet. "By opening the Arab Peninsula to the crusaders," proclaimed bin Laden, "the regime disobeyed and acted against what has been enjoined by the messenger of God."[46] By doing so, claimed bin Laden, Saudi leaders ceased to be Muslims.

For young impoverished Muslims, shuffling to and fro in a world seemingly indifferent to their fate, the simple message of the fundamentalists rings true. The world, they are told, is made up of only two camps, the true believers of Islam and the barbarians. To choose the former camp is to walk in the footsteps of Mohammed and play a vital role in redeeming the world. The very idea of being able to participate in a jihad to change human destiny is electrifying for young people who have been denied even the most basic considerations of life. Here is a different type of participation at work—and not the kind we in the West normally associate with the vote and electoral politics. Our kind of civil democracy enjoys little or no currency for most of the young people of the Middle East. It is not something with which they have any familiarity. In their world, rulers rule. But in Islam, they are told, everyone is equal; everyone counts. By embracing the fundamentalists' cause, they gain instant access to a world in which their dignity is restored, their life has meaning, and their cause is just—a powerful tonic.

Fouad Ajami, in his book *The Arab Predicament*, writes of the allure of the fundamentalist view of participation:

> The fundamentalist call has resonance because it invited men to participate . . . [in] contrast to a political culture that reduces citizens to spectators and asks them to leave things to the rulers. At a time when the future is uncertain, it connects them to a tradition that reduces bewilderment.[47]

The U.S. and the West have been largely ambivalent about encouraging any kind of real democracy in the Gulf. Certainly, the kind of participation that the fundamentalists have in mind is anathema to American policy leaders, but even the more benign kind of democracy that exists in the form of electoral politics has been shunned in policy circles. Fundamentalists are quick to pounce on the American government for entertaining a double standard—that is, promoting democracy at home but supporting dictatorships in the Middle East. There is some truth to the accusations. The U.S. and its Western allies are worried that extending democracy to the Gulf could result in fundamentalists taking power at the ballot box. The prospect of "many Irans" makes American policy leaders nervous. For that reason, the U.S. has cast its lot with existing autocratic regimes like Saudi Arabia, preferring the relative calm that comes with repression to the political instability that might accompany the raucous give-and-take of representative democracy. As one Saudi Arabian professional put it, "Americans want an odorless empire."[48]

In the aftermath of the September 11th attacks on the World Trade Center towers in New York City and the Pentagon in Washington, D.C., it has become clear, even to American officials and an increasing number of rulers in the Persian Gulf, that a certain kind of democracy is coming to the Middle East, whether we like it or not, and that we had better get used to the idea. Just two months after the attacks, Prince Walid bin Talal, a billionaire investor and one of the richest businessmen in the world, as well as a member of the Saudi Arabian ruling family, spoke the unspeakable for the first time. He called for a transformation in Saudi rule that would bring at least a limited form of democracy to the country. Worried over increasing political discontent among Saudi's young people—a majority of the airplane hijackers involved in the September 11th attacks were Saudi youth—the prince said that it was past time to address the previously politically taboo subject of democratic representation: "If people speak more freely and get involved more in the political process," argued the prince, "you can really contain them and make them part of the process."[49]

Prince Walid made it clear that he was not advocating the elimination of the monarchy as the governing body of Saudi Arabia, only suggesting that the existing council to the king be elected rather than appointed. Still, his remarks sent shock waves through the kingdom, leaving many to wonder if the door had opened enough to tolerate free expression of political views in the country.

The most difficult thing for Americans and other Westerners to accept is that democratic expression in the Middle East could propel fundamentalists to power across the region. In fact, it is almost assured to do this. While we prefer to see political parties emerge that are secular, tolerant, and free of gender bias, the reality is that virtually every political movement with any popular support in that part of the world is dedicated to a different agenda: the Islamization of their society and the world. To understand why that is the case, says historian Karen Armstrong, we have to distinguish the Western notion of politics from the Islamic one. In the West, we associate politics with government "of, by, and for the people." In the Islamic world, however, it is not the people who give government its legitimacy, but God. The very idea of people, rather than God, empowering government is considered blasphemy to most Muslims, a kind of direct "usurpation of God's sovereignty."[50]

Democratic reforms allow the Islamists to be voted into office. Once in power, however, Islamic extremists often legislate changes that restrict or curtail the very democratic reforms that allowed them to secure power in the first place. A case in point is Algeria. In 1989, the National Liberation Front (FLN), which had ruled Algeria as a one-party state since it gained independence from France in 1962, established a new constitution that legalized other political parties. In the national elections of 1991, the Islamic Salvation Front (FIS), an Islamic fundamentalist party, won many seats in the National Assembly in the first round of voting. The FIS was expected to go on to win in the second round, giving them control of the government. At that point, the Algerian army intervened and canceled the elections, blocking the FIS bid for power. Had the party succeeded, Algeria would have been restructured along traditional Islamic lines. Religious courts would have replaced civil courts, Islamic dress codes would have been made mandatory, and cultural expression would have been restricted to acceptable Islamic practices. The FIS fought back, and civil war erupted. The FIS leadership was rounded up and arrested, and Algeria was effectively taken over by the military.[51]

Setbacks like the one in Algeria have not slowed the process of "democratization" in the Middle East. Fundamentalist movements throughout the region are organizing themselves into political parties and pushing for electoral reforms. In Kuwait, where there already exists a parliamentary assembly, many observers worry that an Islamized majority could undermine the fragile democratic reforms made thus far. "The democracy has brought Islamists into the Parliament," says Ahmed E. Bishara, a moderate who heads Kuwait's National Democratic Union. But, says Bishara, "you end up with a Parliament that legislates against public freedom."[52] In Kuwait, the Islamists in parliament have voted to separate men and women at the universities, have opposed extending the vote to women, and are pushing to replace civil law with Islamic courts and codes. Islamists like Khalid al-Essa, who heads the conservative Salafi movement in Kuwait, argues that "you can have a democratic organization, but it must take its cues from Islam."[53] For al-Essa and the fundamentalists, Islam is "a boundary condition" for democracy.[54]

In many ways, the fate of the Gulf countries, and the Middle East as a whole, depends on the degree to which the millions of disenfranchised and impoverished youth in the region see the Islamic renewal as their only real hope for making a better future for themselves. If they come to see themselves as Allah's warriors, walking in the righteous path of Mohammed and righting the wrongs of an unjust society, they could become an unstoppable force. If today's rulers come up short on solutions to address the worsening economic situation and continue to govern through a combination of autocratic rule and ruthless put-downs of any kind of opposition voices, it is likely that fundamentalist movements will topple many Middle Eastern regimes and establish radical Islamist states along the lines of Iran over the course of the next ten years. The prospect of young, fiery fundamentalist regimes in power across the Gulf as global oil production peaks is a very real possibility and something for which the rest of the world needs to prepare. Admittedly, there may be little that can be done at this late stage to avoid what many now believe is all but inevitable.

Politicizing Oil

While political leaders and geopolitical strategists worry about the Islamization of oil in the Middle East, the reality is that playing the oil card is a tricky business, as OPEC has learned over the years. In the past, when major oil producers have cut production to keep prices high on world markets, it has generally resulted in a short-term spike in the price per barrel but a long-term decline in both market share and price as oil companies search for cheaper suppliers. Certainly this was the case in the 1970s and 1980s. Catching the world off-guard in the 1970s, Gulf producers were able to capitalize on their oil-embargo and production cuts, and they reaped windfall profits for nearly a decade. By the 1980s, however, oil companies had ramped up alternative sources, forcing Middle East producers to drop their prices and share a smaller part of the world oil market.

The new circumstances that will accompany the peak in global oil production, however, will likely create a very different dynamic. This time around, other sources of cheap crude oil will not be available in the volume needed to meet any deficit that might occur if Gulf countries curtail production and raise prices. As we have seen, geologists might disagree about exactly when global oil production is likely to peak, but they agree that two-thirds of the remaining global oil reserves, post-production peak, lie in the Middle East. So whether the existing autocratic regimes in the Middle East stay in power or the fundamentalists overthrow them, oil power is going to shift back to the region in the years ahead. And when it does, whoever is in power will be in a position to dictate the terms on world oil markets, simply because there is no other place to go for abundant crude oil.

The only real difference, then, between the old guard remaining in power and the new militants seizing power will be whether Middle East governments will use their new position of dominance to exact strictly commercial gains or whether political considerations will also affect the flow of oil. From a purely expedient perspective, keeping the oil flowing—albeit at a dramatically increased price—makes all the sense in the world, since it brings in more revenue for the producing countries. On the other hand, it is not impossible to imagine situations in which fundamentalist regimes decide to turn off the spigot, at least for a short period of time, and hold the world hostage, in hopes of exacting political concessions of various kinds.

On April 8, 2002, President Saddam Hussein of Iraq attempted to do just that, announcing that his country was suspending the export of all its oil for thirty days to protest the Israeli incursion into "Palestinian territories." Iraq is the sixth-largest oil supplier to the U.S., providing 9 percent of U.S. oil imports. Iraq exports a total of two million barrels a day, which accounts for 4 percent of all the oil traded on world markets.[55] Earlier, Iran's foreign minister, Kamal Kharrazi, announced that his country would also support an oil embargo—Iran is the second-largest OPEC oil producer—if other Arab oil-producing countries agreed to go along with the Iraqis. Libya said that it would also support an oil embargo if other Arab countries agreed to do so as well.[56]

The Saudis—OPEC's largest producer—say that they oppose using oil as a weapon and point out that it would only hurt the Arab oil-producing countries who can ill afford a drop in oil revenue. Many Persian Gulf countries rely on oil revenue for more than two-thirds of government income.[57]

For now, the prevailing wisdom is that there is "a very low possibility" that Arab oil producers would cut oil supplies to exert pressure on Israel and the U.S. Still, things could change quickly, and the consequences could be profound and far-reaching.[58]

In the short run, Russia, Norway, Canada, Mexico, and other non-OPEC oil producers could crank up production to keep oil flowing, but even the partial withdrawal of OPEC oil from world markets could send the price of oil soaring to as much as $50 or more per barrel. Were that to happen, it would trigger a devastating downturn in the global economy.[59]

The threat of oil embargoes notwithstanding, oil prices are destined to rise once global oil production peaks, and the Middle East oil-producing countries will be the economic and political beneficiaries. The transfer of wealth from industrial powers like the United States to Gulf producers will be staggering. The heavy reliance on foreign oil could result in a $100-billion-per-year outflow of cash from the U.S., seriously exacerbating America's already large trade deficit. For the Gulf, the new reality could mean a $160 billion increase in yearly revenue by 2010.[60]

That kind of incredible wealth, generated in so short a period of time, is going to deeply affect Arab culture as well as regional politics in the Middle East. The one-way flow of wealth from the rest of the world to the Persian Gulf is likely to further heighten geopolitical tensions between the Muslim countries and the West, leading to more-open conflict and a long and protracted struggle between the two forces.

6

A GLOBAL MELTDOWN

The consequences of militant takeovers in countries throughout the Middle East, at the very time that global oil production is likely to peak, will be felt far beyond the gas pump. Governments and energy companies are already scrambling to diversify their energy portfolios. In the immediate future, greater emphasis is being put on the exploration and development of natural gas. Unfortunately, by over-relying on natural gas to take up the slack when global oil production peaks, energy companies and electrical utilities may be creating a second energy crisis that will come fast on the heels of the first energy crisis with crude oil and that could prove devastating to the global economy.

Others are promoting the increased use of coal as well as the tapping of heavy oils, tar sand, and shale, all "dirty," CO_2-emitting fossil fuels that could exacerbate the already dangerous rise in the Earth's temperatures.

While the natural-gas route could lead to a dead end, the switch to dirtier fuels could lead to a head-on clash between geopolitics and the protection of the biosphere, posing unprecedented risks for both human civilization and the Earth itself.

Making Do with Natural Gas

The good news about natural gas is that it is a less-polluting form of fossil fuel than either oil or coal. Oil produces one-third more CO_2 than gas per equivalent unit of energy produced, while coal produces two-thirds more CO_2.[1] For years, environmentalists and policy leaders have been urging industry to make the transition to natural gas—the cleaner fuel—in hopes of reducing dangerous greenhouse-gas emissions.

The bad news is that new studies suggest that global production of natural gas is likely to peak soon after global oil production, with some analysts predicting a peak as early as 2020. Dutch Shell says that "[natural-gas] scarcity could occur as early as 2025."[2] Compounding the problem, much of the remaining natural-gas reserves—more than 40 percent of the total—are in the Middle East.[3] With reserves already nearing peak in North America, we will find ourselves—along with many other regions of the world—increasingly at the mercy of the Middle East swing producers and Russia by the third decade of the century, a situation that will further narrow our energy options and put the future of the global economy at risk.

Gas is often found in regions where there are large oil or coal deposits, but it is also found in places where there is no oil or coal at all. In the past, natural gas has been used for space heating or as boiler fuel. Today, it is fast becoming the fuel of choice for the generation of electricity. It is also increasingly being used as a fuel for transportation. New technological breakthroughs in converting gas to liquid have reduced costs and brought liquid gas to a range that is competitive with gasoline for some transportation-related activities.

Approximately 14 percent of U.S. natural-gas consumption is currently used to generate electricity.[4] That figure is going to go up dramatically over the course of the next decade. There are 272 gas-fired power plants currently under construction or on order in the U.S. They are expected to come on line in the next ten years, making the North American electricity grid virtually dependent on natural gas.[5] Utility companies favor gas-fired generators over coal, oil, or nuclear because they are less capital intensive, have shorter construction lead times, are more efficient, and have lower emissions.[6]

The U.S. Department of Energy projects natural-gas consumption to increase from its 1999 production output of 21.4 trillion cubic feet to between 32.2 trillion and 36.1 trillion cubic feet by the year 2020. Nearly 57 percent of the increase will be used to power the new generation of gas-fired utilities coming on line, and the rest will be used to meet the increasing demand expected in the residential, commercial, industrial, and transportation sectors.[7]

New computer-model studies surveying the remaining natural-gas reserves in North America suggest that the enthusiasm for natural gas, as an alternative to scarce oil, may be misplaced. According to a study conducted by electrical engineer Richard Duncan, U.S. production peaked in 1971 at 22 trillion cubic feet (Tcf)—only a year after U.S. crude-oil production peaked. From 1971 to 1999, U.S. gas production declined by 0.50 percent per year. Duncan predicts a secondary U.S. peak in 2007 at 20.1 Tcf, followed by a steady decline in production of 1.5 percent a year until 2040.[8] With the domestic demand for natural gas expected to increase by 62 percent between now and 2020—natural-gas demand for electricity alone will triple during this time period—remaining reserves will be quickly depleted.[9] Similarly, Canada's natural-gas production is expected to peak in 2005 and will begin to decline by an average of 4.3 percent a year for the next thirty-five years.[10] Mexico's production is forecasted to peak by 2011 at 1.5 Tcf and fall by 2.7 percent a year for the next twenty-nine years.[11]

North America, as a whole, will peak in natural-gas production as early as 2007 at 28.5 Tcf. From then to 2040, production will fall by 51 percent, with an average decline of 2.1 percent a year.[12] These kinds of figures worry some experts who wonder about the consequences of running low on North American gas reserves right at the time when much of the North American electrical grid is powered by natural gas. Duncan, in his keynote address to the Geological Society of America's Summit in 2000, predicted rolling electrical blackouts in the U.S., beginning around 2012, due to a shortage of readily available natural gas.[13]

We could import natural gas from abroad—which is likely—but it will not come cheap. Most natural gas is transported over land by pipeline. It can be cooled to a liquid and shipped in cryogenic tanks by sea, but at great additional expense. The refrigeration costs in keeping the gas cooled are high. Still, with the prices of both oil and natural gas expected to rise, importing natural gas will likely be cost-competitive by the end of the decade. Natural gas from Algeria has already been shipped to the United States, and Japan receives tanker shipments of natural gas from Alaska and the Middle East.[14]

Importing natural gas will only buy a few years of additional time, say a growing number of geologists. C. J. Campbell believes that the world peak in gas production could come as early as 2020.[15] After that, the global economy will increasingly rely on the Middle East and the former Soviet Union for its gas needs, and at dramatically escalating prices. Iran, a recently Islamized nation, has 16 percent of the world's gas reserves, and Qatar and the United Arab Emirates have another 10 percent, making all three countries key players in the future global energy game.[16] Again, as with oil, the Middle East becomes the swing producer in a world running low on natural gas as well as cheap oil.

The global struggle to secure crude oil and natural gas is going to intensify as we turn the corner into the second decade of the 21st century. This time, the developing world—especially China and India—are going to be as dependent on these fossil fuels as are the leading industrial nations, making the competition for both fuels a truly global phenomenon. Steadily rising prices are likely to create havoc in the global economy, increasing the prospects of hyperinflation, recession, and even depression. As oil and gas become scarce, every other economic activity downstream is affected. The days of cheap energy are passing—and with that passage it is unrealistic to expect the kind of economic growth that was experienced during the 20th century. If energy shortages become so acute that the electrical grid itself is compromised, resulting in rolling brownouts and blackouts across the globe, then we could potentially see the collapse of the entire infrastructure that supports a complex global economy and human society. Were this to happen, the future might look very different from anything we are familiar with or can even model or imagine.

Heavy Oils and Rising Temperatures

Geologists and economists remind us that diminishing reserves of cheap crude oil and natural gas don't mean that the world is running short on fossil fuels. There is still plenty of coal, tar sand, heavy oil, and oil shale out there ready to be explored, mined, and processed, once the economics of the energy market make them cost-competitive. These so-called unconventional sources of oil are looked to as a remedy of last resort, and industry boosters are already touting the possibilities, both in trade circles and in the public arena. The Paris-based International Energy Agency of the OECD, in its World Energy Outlook, predicts that as the price of crude oil increases per barrel, the oil-consuming countries will begin to look to these unconventional sources.[17] That process could be hurried up if Middle East producers raise prices precipitously and/or dramatically restrict production for either commercial or political reasons.

The shift to unconventional sources of oil brings with it a heavy price, both to society and the planet. These other fossil fuels are dirtier than either oil or natural gas. In a world desperate to keep the power on and cars running, many may feel there is little choice but to sacrifice the long-term interest of the biosphere for the short-term needs of the economy by burning increasing amounts of unconventional fuels. Fuels extracted from coal, heavy oil, and tar sand would increase the emission of carbon dioxide into the atmosphere, forcing global temperatures to rise even higher than is currently projected by the international scientific community. It should be recalled that the existing global-warming models assume the continued high-volume use of conventional oil and natural gas well into the middle decades of the 21st century. Foreshortening that timetable by several decades and increasing the use of unconventional sources of oil will affect both the rate and timing of global warming.

How much unconventional oil is the world going to have to burn just to keep the global economy afloat? Most conservative forecasts predict a doubling of energy use between the years 2000 and 2040, a tripling by 2070, and a quadrupling by 2100 to meet the increased needs of a growing human population.[18] This means a tripling of carbon-dioxide emissions from six billion tons of carbon per year in 2000 to twenty billion tons by 2100.[19] If the transition to unconventional oils were to occur by 2015, rather than by 2050 as previously projected, and continue to increase as oil and gas become more scarce and demand for energy climbs, CO_2 emissions would grow concomitantly with potentially devastating effects on climate.

The U.S. has the largest coal reserves in the world. Large coal reserves also exist in the former Soviet Union—about equivalent to those in the United States. China, India, Germany, Australia, and South Africa each contains between 6 and 12 percent of the world's reserves.[20] The U.S. uses coal for 23 percent of its total primary energy requirements.[21] Currently, 55 percent of the electricity generated in the U.S. comes from coal-burning power plants.[22] Coal-industry analysts hope that, as the costs of oil and natural gas rise, coal liquefaction for synthetic fuel will become an increasingly attractive, cost-competitive option.

Whereas China and India are expected to count for more than 92 percent of the total expected increase in the use of coal between now and 2020, an early peak in global oil production could put coal back into the energy mix in a major way in many industrial nations.[23] Coal is already enjoying a renaissance of sorts in the United States. The coal industry likes to remind Americans that the U.S. is sitting on the largest coal deposits in the world and that, if they were properly utilized, we could wean ourselves away from reliance on foreign oil, especially from the Middle East, and become energy-independent. In the aftermath of the September 11th attacks, the coal industry picked up considerable support among policymakers in Washington who are anxious to reduce American dependence on Persian Gulf oil.

The Bush Administration supported legislation that would provide several billion dollars in research grants, tax credits, and subsidies for the development of what the industry euphemistically calls "clean coal technology."[24] (Critics charge that the new technologies do little to reduce CO_2 emissions.) With the new interest being generated in coal, industry analysts are projecting a growth in coal consumption from between 1 and 2 percent per year to between 3 and 4 percent per year.[25]

Geologist Craig Hatfield says that even if one accepts the rather "fanciful" claim that the U.S. has a 300-year supply of producible coal at existing consumption rates—new revisions in reserve figures show fewer reserves—if consumption were to increase to 4 percent, as the industry is projecting, the reserves would last only sixty-four years. And, says Hatfield, this assumes that coal would continue to be produced at the same rate until depletion of the resource.[26] Equally important, notes Hatfield, is how little liquid fuel can be produced from a ton of coal—only 5.5 barrels of liquid per ton. To replace just 10 percent of total global oil consumption in 1996, for example, would require liquefying an amount of coal half as great as the total mined in the entire country during the past ten years—namely, a 50 percent increase in the rate of coal extraction.[27]

Tar sand and heavy oil are also being looked to as substitutes for crude oil. Geologists estimate that there may be as much as 300 billion barrels of recoverable oil in tar sand in Northern Alberta, Canada. Venezuela is also believed to contain 300 billion barrels of recoverable extra heavy oil. Canada's and Venezuela's combined potentially recoverable reserves are double the conventional reserves of oil in Saudi Arabia and are equivalent to the total recoverable reserves of conventional oil in the entire Middle East.[28]

Large tar-sand deposits have also been found in Estonia, Australia, Brazil, the U.S., and China.[29] The United Nations Institute for Training and Research (UNITAR) Centre for Heavy Crude and Tar Sands estimates that the amount of recoverable heavy oil in the world is roughly equivalent to one-third of the world's total oil and gas reserves, making it a potentially significant player in the global energy mix.[30] In a recent report, the Centre concluded that "the peaking of conventional oil, combined with physical shortages and crude price escalation [sometime in the second decade of the coming century], will mark the beginning of commercial production from the world's huge deposits of tar sands and extra heavy oil."[31]

Up to now, heavy oil has been slow to develop because of its high viscosity and levels of sulfur, metals, and nitrogen, which make it expensive to produce, transport, and refine.[32] At present, heavy oil makes up only about 3.5 percent of total world oil production.[33] Still, countries like Venezuela and Canada are making significant investments in heavy-oil and tar-sand exploration and production in hopes of being able to move swiftly into the energy vacuum once global oil production peaks. Venezuela forecasts that heavy oil will make up more than 40 percent of its overall oil production by 2010. In Canada, industry observers expect tar sand to comprise 75 percent of domestic oil production by the end of the decade.[34] Together, Canada and Venezuela expect to produce one million barrels per day by 2010.[35]

Syncrude Canada, which operates the largest tar-sand oil production facility in the world, is already producing more than 13 percent of Canada's total oil requirements.[36] Shell Canada, Chevron, and Western Oil Sands have recently begun construction of a state-of-the-art $2.35-billion tar sand facility in northwestern Canada, banking on tar sand as the fossil-fuel resource of the future.[37]

Despite the enthusiasm, significant obstacles remain in the way of making tar sand commercially competitive with crude oil and natural gas. To begin with, it takes two tons of tar sand to produce just one barrel of oil. Tons of rock have to be hauled from the open pits and crushed into smaller particles. Then the oil has to be extracted from the tar sand either by hot-water processes, solvents, or thermal means. The extracted oil has to be further refined into fuel oil.[38] The whole process is expensive. Syncrude Canada spends about $12 to produce a barrel of oil. By contrast, the Saudis spend only $1 for every barrel of oil produced.[39] Syncrude says that it can make a decent profit if world oil prices hover at $21 per barrel. Industry analysts, however, say that the threshold price for making tar sand commercially viable is probably $25 per barrel. One-half of the technically recoverable reserves could be potentially produced as liquid fuel, say economists, if the price of oil were to climb to $45 per barrel.[40] Although $45 per barrel seems very high today, when global oil production peaks and crude oil becomes increasingly scarce, the higher price could become the norm. At least that is what the investors in tar sand production are gambling on.

Mining and refining tar sand is also very dirty. Syncrude Canada emits 240 tons of sulfur dioxide a day, twenty-five times the amount emitted by a conventional refinery producing the same amount of oil in Texas.[41]

Mining and processing tar sand require a large amount of water, meaning that less water is available for other purposes, including farming, residential, and commercial uses. Moreover, the water must be heated. One recent study forecasts that by 2010 nearly 25 percent of Alberta's produced gas will be used just to heat the water to melt the bitumen.[42] Water shortage has become a lightning-rod issue among environmentalists in recent years and is likely to intensify as more water is used to process tar sand.

Of major concern is the tailings slurry from the extraction process. The tailings contain hydrocarbons, inorganic salts, and heavy metals. The sludge, which is gelatinous in composition, cannot be reclaimed. The sheer volume of sludge accumulating in tailings ponds in Canada's tar sand fields is disquieting. Suncor and Syncrude, two main producers, will have tailings ponds that exceed one billion cubic meters by the year 2020. Because the sludge can take more than a century to consolidate into a form that can be safely reclaimed, the tailings ponds cannot be consolidated to ensure against erosion and breaching. Migration of sludge pollutants into soil and groundwater could pose serious long-term environmental problems.[43]

If soil and groundwater contamination were the only risks posed to the environment by mining and refining tar sand and heavy oil, the damage, which is potentially significant, would at least be localized in its effects. The larger problem in replacing a diminishing supply of cheap crude oil with tar sand and heavy oil is the increase in CO_2 emissions. The lower the fuel-conversion efficiency, the greater the emission of CO_2. For example, producing synthetic oil from oil shale results in 39 percent more CO_2 emissions than from producing crude oil. Producing synthetic oil from coal results in 72 percent more CO_2 emissions than from producing crude.[44] Similarly, converting tar sand and heavy oil to synthetic fuels results in significantly higher emissions.

The potential significance of this simple thermodynamic fact can't be overstated, especially when we consider how dependent the world may become on heavy oil and tar sand in the years ahead. A study done more than a decade ago by Chevron forecast that, by the mid-21st century, heavy oil and bitumen could make up more than half of the world's energy supply.[45] If heavy oil, tar sand, and coal are increasingly used to produce synthetic liquid fuel to make up for a diminishing supply of cheap crude oil, the long-term impact on the world's climate could be even worse than anyone is presently predicting.

The Entropy Bill for the Industrial Age

Global warming represents the dark side of the commercial ledger for the Industrial Age. For the past several hundred years, and especially in the 20th century, human beings burned massive amounts of "stored sun" in the forms of coal, oil, and natural gas to produce the energy that made an industrial way of life possible. That spent energy has accumulated in the Earth's atmosphere and has begun to adversely affect the climate of the planet and the workings of its many ecosystems. Like the other civilizations that have come before, industrial society is approaching the final stage of its own energy regime, during which the costs of absorbing the accumulated wastes of all the energy already spent are becoming as significant an economic factor as the net value of the available energy still being produced and consumed.

If we were to measure human accomplishments in terms of the sheer impact our activities have had on the life of the planet we inhabit, we would sadly have to conclude that global warming is humankind's most significant accomplishment to date, albeit a negative one. We have literally affected the very biochemistry of the Earth, and we have done so in less than a century. When future generations look back at this period, tens of thousands of years from now, the only historical legacy we will have left them in the geological record is a qualitative change in the Earth's climate.

The heating up of the Earth is the result of a growing accumulation of gases in the atmosphere that are blocking heat from escaping the planet. This "greenhouse" phenomenon begins when solar radiation enters the Earth's atmosphere, hits the surface of the Earth, and is transformed into infrared energy and heat. The heat rises and bombards carbon dioxide and other gases in the Earth's atmosphere, forcing the gaseous molecules to vibrate. The gas molecules act as reflectors, sending some of the heat back to the surface, creating a warming effect. Carbon dioxide, methane, and other greenhouse gases provide an atmospheric blanket that allows enough of the heat generated by the sun's radiation to stay on Earth to provide the right conditions for the flourishing of life. For the 10,000 years leading up to the Industrial Age, the balance of greenhouse gases was relatively stable, so that the temperature on the planet remained within a narrow range. The burning of massive amounts of coal, then oil and natural gas in the 19th and 20th centuries, changed the equation.

The CO_2 content in the atmosphere today is approximately 31 percent greater than it was in 1750, at the onset of the fossil-fuel era. That concentration, according to the UN Intergovernmental Panel on Climate Change (IPCC), has not been exceeded in the past 420,000 years and likely not in the last 20 million years. The rate of increase in CO_2 concentration, according to the panel, is without precedent in the past 20,000 years. Nearly 75 percent of the increase in CO_2 concentrations in the past twenty years is attributable to the burning of fossil fuels. The remainder is the result of deforestation and land-use changes, both of which release CO_2 into the atmosphere. While the land and ocean absorb half of the increase in CO_2 emissions, the rest migrates to the atmosphere.[46]

Methane, another greenhouse gas, has also contributed to the thickening of the greenhouse blanket and the trapping of greater heat on the Earth's surface. More than half of the methane emissions are human-induced and include emissions from rice paddies, landfills, and animal flatulence. Methane has increased in concentration in the atmosphere by 151 percent since 1750. Like the CO_2 concentrations, the amount of methane in the atmosphere is unprecedented in the geological record of the past 420,000 years.[47]

Nitrous oxide, the third major greenhouse gas, has increased in concentration by 17 percent since 1750. Nearly one-third of the increase is human-induced and includes emissions from cattle feedlots, the chemical industry, and the heavy use of chemical-based fertilizers in agricultural soil.[48]

The increase in carbon dioxide is responsible for more than 70 percent of the global warming effects, while methane makes up 24 percent and nitrous oxide only 6 percent of the effects.[49]

The first concern about global warming came in the form of a scientific paper published in 1957 by Roger Revelle and Hans Suess of the Scripps Institute of Oceanography in California. The two scientists warned that human industrial and agricultural activity was resulting in a dangerous buildup of CO_2 in the atmosphere, with untold consequences to the Earth's temperature.[50]

Interest in the subject picked up in the 1970s and '80s, in part because of the oil crisis and also because of increasing concern for the environment. Scientists from a range of fields began conducting studies, making forecasts, and bickering with one another over the facts and figures and the interpretation of data. In the late 1980s, in an attempt to make sense of all the different spins on climate change, the United Nations assembled a group consisting of hundreds of the world's leading scientists from more than a dozen different disciplines and asked them to undertake a rigorous study of the question. After more than a decade of work and several interim reports, the UN Intergovernmental Panel on Climate Change issued a voluminous consensus report in January 2001. Their findings are sobering.

The IPCC scientists say that the global average surface temperature increased by 1.08 ± 0.36°F over the course of the 20th century. Moreover, the increase is likely to have been the largest during any century in the past 1,000 years.[51] And, according to the scientists' computer models, the global average surface temperature is forecast to increase by between 2.52 and 10.44°F by the year 2100.[52]

To put all this in perspective, the scientists say that the projected temperature increases are "without precedent during at least the last

10,000 years based on palaeoclimate data."[53] In other words, if the projection holds up, the change in temperature forecast for the next 100 years will likely be larger than any climate change on Earth in more than ten millennia. The impacts on the Earth's biosphere are going to be of a qualitative kind. Recall that a 9°F increase in temperature between the last Ice Age and today resulted in the uncovering of a large area of the Northern Hemisphere of the planet from beneath thousands of feet of ice.[54]

The anecdotal evidence that rising temperatures are adversely affecting the Earth's ecosystems is appearing everywhere. Sometimes small changes are enough to raise large concerns. Recently, researchers reported that the Edith's Checkerspot butterfly of western North America had migrated more than sixty miles north of its traditional range.[55] The Pacific salmon populations fell sharply in 1997 and 1998 as ocean temperatures climbed by more than 6°F.[56] Mount Kilimanjaro, Africa's tallest peak, has lost 75 percent of its ice cap over the course of the past century and will likely be ice-free within less than fifteen years.[57] Polar bears in the Hudson Bay are having fewer cubs, and scientists speculate that this may be due to the earlier-spring ice melt.[58] Dengue fever and other tropical diseases are invading the southern United States for the first time.[59]

Scientists forecast that a temperature rise of 2.52 to 10.44°F over the course of the coming century could have devastating long-term effects on the Earth's ecosystems, including the melting of glaciers and the Arctic polar cap, a rise in seawater levels, increased precipitation and storms and more violent weather patterns, destabilization and loss of habitats, northward migration of ecosystems, contamination of freshwater by salt water, massive forest dieback, accelerated species extinction, and increased droughts. The IPCC report also warns of adverse impacts on human settlements, including the submerging of island nations and low-lying countries, diminishing crop yields (especially in the Southern Hemisphere), and the spread of tropical diseases northward into previously temperate zones.

The scope and dimension of the changes taking place are breathtaking. Consider the Mendenhall glacier in southeastern Alaska. This Alaskan landmark near Juneau attracts more than 300,000 visitors a year and is one of the world's most-watched glaciers. The glacier, say climatologists, is shrinking. In the summer of 2000, it retreated more than 100 meters, exposing land that had been under ice for centuries. Over the past seventy years the glacier has shrunk by a kilometer. Keith Echelmeyer, a geologist who has been studying the fast-changing glacial landscape in Alaska, says that, based on the current rate of ice melt, the glacier may disappear before the end of the 21st century. Other glaciers in the region are experiencing similar losses, and many will not survive the next half century, says Echelmeyer.[60]

Satellite data show a 10 percent decrease in snow cover since the 1960s in the middle and high latitudes of the Northern Hemisphere. Other tracking data show a two-week reduction in river and lake ice duration in the same regions.[61] Arctic sea ice has also declined in thickness by 40 percent or more between late summer and early fall in recent decades.[62] These decreases are projected to accelerate during the course of the present century.

The IPCC also reports a disturbing rise in seawater levels around the world. The newest data show a rise in global mean sea level of between 0.1 and 0.2 meters during the 20th century. More alarming, the computer models suggest that the levels will rise by an additional 0.09 to 0.88 meters between now and the end of the century.[63] While some of the rise will result from the melting of glaciers and the ice caps, much of the increase, say forecasters, will be due to thermal expansion from the heating up of the oceans. Over the next 500 years, seawater could rise as much as seven to thirteen meters.[64] Because it takes nearly a thousand years for warming in the atmosphere to reach the ocean floor, the increase in the Earth's surface temperatures in the next century will continue to raise sea levels for centuries thereafter.[65]

Of particular concern to climatologists is the Greenland ice sheet, which is beginning to melt. If temperatures in Greenland rise above 36.86°F, the Greenland ice sheet will disappear, causing sea levels to rise by as much as seven meters.[66] If sea levels were to rise by ten meters or more over the next 1,000 years because of the temperature changes already in motion, the inundation could result in loss of landmass comparable in size to the United States. Florida and many cities on the eastern seaboard of the U.S. would be under water. More than a billion people would have to relocate, and much of the best farmland in the world would be submerged.[67]

Hardest hit would be the many island nations, like the Maldives in the Indian Ocean and the Marshall Islands of the Pacific. Just a half-meter rise in sea level would damage their coastlines and reduce their freshwater reserves. A three-meter rise would submerge these islands completely.[68]

Half of the human race inhabits coastal zones, making much of the population vulnerable to rising sea levels. In low-lying countries like Bangladesh that are already experiencing increasing flooding each year, a seawater rise of one meter could eliminate 7 percent of the land, affecting six million people. People living along the Nile delta in Egypt could face similar hardships.[69]

The rise in Earth's temperatures is also going to affect precipitation patterns. The IPCC notes that precipitation increased by 0.5 to 1 percent per decade in the 20th century over the middle and higher latitudes of the Northern Hemisphere continents.[70] Heavy rainfall has increased by 2 to 4 percent in the same regions, and some scientists predict, based on the most recent computer models, that violent storms will intensify in the middle latitudes.[71] The IPCC projects that more intense coastal storm surges could increase by several-fold the number of people affected by flooding.[72]

Imagine central Canada having the climate of present-day central Illinois, or New York City having the climate of Miami Beach, Florida.[73] While human populations will be able to trek north, plants and animals may not be able to migrate quickly enough to keep pace with their own temperature ranges. Many ecosystems, unable to adjust, will die out or be replaced by new regimes.

Sir John Houghton, co-chairman of the IPCC panel and chairman of the United Kingdom's Royal Commission on Environmental Pollution, points out that microorganisms, plants, and animals throughout the world have evolved and adapted to specific climatic zones. "Changes in climate," says Houghton, "alter the sustainability of a region for different species and change their competitiveness within an ecosystem, so that even relatively small changes in climate will lead, over time, to large changes in the composition of an ecosystem."[74] Of course, the temperature changes that Houghton has in mind take place over the course of thousands of years. Global warming, however, threatens to alter climatic conditions in less than a century. "Most ecosystems cannot respond or migrate that fast," says Houghton. When ecosystems and climates become increasingly mismatched, the ecosystems become more prone to disease, attacks by pests, and other assaults.[75]

Because trees, in particular, live so long and take so long to reproduce, forests are particularly vulnerable to disruption, destabilization, and dieback when the temperature changes rapidly. Since trees cover one-fourth of the land surface of the Earth and are home to many of the world's creatures, any change in global temperature would have a significant effect on the forest ecosystems of the planet.[76]

Researchers working in Costa Rica over the past sixteen years have recorded a continuous decline in the growth rate of trees in the rain forests as surface temperatures there have steadily increased. Researcher Deborah A. Clark of the University of Missouri reports that "tropical trees are being increasingly stressed through higher nighttime temperatures."[77] The higher nighttime temperatures also force the trees to respire more and release more CO_2. These findings worry scientists. Researchers studying the effects of temperature rise on declining tree growth observe that tropical rain forests absorb as much as one-third of all the CO_2 taken out of the atmosphere by photosynthesis on land.[78] If trees continue to release more and more CO_2 as a by-product of respiration, in effect shifting the balance between intake and release of CO_2, the amount of CO_2 added to the atmosphere could significantly increase global temperature above the current forecasts. Peter Cox, of the U.K. Meteorological Office in Bracknell, says that in the coming decades we could see massive forest dieback in the Amazon as a result of increasing heat stress, and the release of billions of tons of CO_2 into the atmosphere. The virtual collapse and die-out of the Earth's largest remaining sink of CO_2 could push global temperatures up by an additional 3.6°F over the course of the century, according to John Mitchell, also of the U.K. Meteorological Office.[79]

Global temperature rise does not mean only doom and gloom,

point out the climatologists. The rise in global temperatures is already having an ameliorative effect on tree growth in the colder latitudes. Satellite data collected by NASA show a greening trend in parts of Canada, the northern U.S., northern Europe, Russia, and central Asia. Earlier springs are extending the growing season by as much as twelve days in North America and eighteen days in Europe and Asia, resulting in an 8 to 12 percent greener environment.[80] Whether the growth in vegetation in the northern latitudes can make up for the loss of growth in the tropical regions, in terms of balancing the uptake and release of CO_2, is still very much unknown.

Climate change is also going to have a significant effect on the "distributions, population sizes, population density and behavior of wildlife," say the IPCC authors. For example, the warming of oceans will mean a shrinking in habitats for cold-water fish and an expansion of range for warm-water fish.[81] The ocean's great coral reefs, among the most fragile living systems, could disappear altogether in less than fifty years from extensive bleaching caused by rising ocean temperatures, according to a report published by the Marine Biological Station of the universities of Glasgow and London. The authors of the report say that, as sea temperatures rise, the microscopic plants that the tiny coral animals feed on die, and the destabilization of that delicate balance, in turn, causes the coral to die. That process is already well under way, affecting coral reefs around the world. Rupert Ormand, of the Marine Biological Station, says that "within fifty years there will be very little left of corals."[82]

Intensification of droughts also figures prominently in all of the newest computer studies. A study by the Tyndall Centre for Climate Change Research at the University of East Anglia in Norwich, U.K., predicts that some countries could warm up by as much as 9°F in the coming decades. Especially at risk are a number of Asian countries, including Kazakhstan, Uzbekistan, Tajikistan, Afghanistan, and Iran. African countries, including Ethiopia, Sierra Leone, and Tanzania, are also likely to see a worsening of drought conditions that have already devastated their lands.[83]

Nowhere will the impacts of global warming be more keenly felt than in agricultural production. Bill Easterling, professor of geography and agronomy at Pennsylvania State University and a lead author of the IPCC report, says that if the temperature increase in the northern latitudes was slight—less than a few degrees—crop yields would likely benefit. Beyond a 3°F warming, however, there would be a dramatic turning point in places like the U.S., where the "crop yields would start to decline rapidly."[84] In the tropics and the Southern Hemisphere, where crops are already at the limit of heat tolerance, the effects of global warming will come sooner and be more pronounced and long-lasting. The developing nations of the Southern Hemisphere, already troubled by growing populations, will likely find it increasingly difficult to feed their own people, let alone grow food for exchange in world markets.[85]

Perhaps the biggest unknowable is how rising temperatures will affect the migration of diseases into previously unexposed regions of the world. The authors of the IPCC report say that, based on the results of their predictive models, "there is medium to high confidence" of a "net increase in the geographic range of potential transmission of malaria and dengue."[86] Malaria flourishes best in warm, humid climates and is spread by mosquitoes. More than 350 million people per year, mostly in the Southern Hemisphere, are infected, and two million people die annually of the disease.[87] Other diseases, including viral encephalitis, are likely to spread as their temperature range moves north.

A Worse-Case Scenario

It should be noted that the current projections about global warming and its ecological, economic, and social impacts are based on the assumption of a steady upward climb in temperatures, more or less evenly distributed over the course of the 21st century. But that assumption may be faulty. The U.S. National Academy of Sciences (NAS) issued a truly frightening report in 2002, which raised the possibility that Earth temperatures could rise suddenly and drastically in just a few years' time, creating a new climatic regime virtually overnight.

The authors of the report point out that abrupt changes in climate, whose effects are long-lasting, have occurred repeatedly in the last 100,000 years. For example, at the end of the Younger Dryas interval about 11,500 years ago, "global climate shifted dramatically, in many regions by about one-third to one-half the difference between ice-age and modern conditions, with much of the change occurring over a few years."[88]

According to the study, "an abrupt climate change occurs when the climate system is forced to cross some threshold, triggering a transition to a new state at a rate determined by the climate system itself and faster than the cause."[89] Moreover, the paleoclimatic record shows that "the most dramatic shifts in climate have occurred when factors controlling the climate system were changing."[90] Given the fact that human activity—especially the burning of fossil fuels—is expected to double the CO_2 content emitted into the atmosphere in the current century, the conditions could be ripe for an abrupt change in climate around the world, perhaps in only a few years. The authors of the NAS report write:

> Current trends along with forecasts for the next century indicate that the climate averages and variabilities likely will reach levels not seen in instrumental records or in recent geological history. These trends have the potential to push the climate systems through a threshold to a new climate state.[91]

What is really unnerving is that it may take only a slight deviation in boundary conditions or a small random fluctuation somewhere in the system "to excite large changes . . . when the system is close to a threshold," says the NAS committee.[92]

An abrupt change in climate, like the kind that occurred during the Younger-Dryas interval, could prove catastrophic for ecosystems and species around the world. During that particular period, for instance, spruce, fir, and paper birch trees experienced extinction in southern New England in less than fifty years. The extinction of horses, mastodons, mammoths, and saber-toothed tigers in North America was greater at that time than in any other extinction event in millions of years.[93]

The committee lays out a potentially nightmarish scenario in which random triggering events take the climate across the threshold into a new regime, causing widespread havoc and destruction. Ecosystems could collapse suddenly, their forests decimated in vast fires and their grasslands dying out and turning into dust bowls. Wildlife could disappear, and waterborne diseases like cholera and vector-borne diseases like malaria, dengue fever, and yellow fever could spread uncontrollably beyond host ranges, threatening human health around the world.[94]

The NAS concludes its report with a dire warning:

> On the basis of the inference from the paleoclimatic record, it is possible that the projected change will occur not through gradual evolution proportional to greenhouse gas concentrations, but through abrupt and persistent regime shifts affecting subcontinental or larger regions . . . denying the likelihood or downplaying the relevance of past abrupt changes could be costly.[95]

IT HAS BECOME increasingly apparent in recent years that the fossil-fuel age has a Janus face. The extraordinary benefits of utilizing coal, oil, and natural gas are too many to recite. Suffice it to say that eight to ten generations of human beings—those living between the time of the first widespread application of coal and steam technology and today—in Europe, North America, Japan, and elsewhere reaped unprecedented gains from the exploitation of these unique, nonrenewable resources. We exhumed the organic remains of an earlier geological era and basked in a material cornucopia made possible by the energy we took up. Now, a number of forces are converging to create a watershed event of great historic proportions. We have reached the critical moment that many great civilizations in the past have confronted, some successfully, others not. That is the point at which the energy used to maintain the workings of the civilization becomes more scarce and expensive, and the accumulated waste and externalities built up from past activity become more costly to absorb. When that juncture is crossed, countries experience a lessening of the energy flow-through, a slowing of the performance of the many subsystems that make up the society, and a weakening of the institutional, economic, and social fabric, making the overall operational structure more vulnerable to external threats and internal collapse.

The choices that a civilization makes at the "turning point" in its existing energy regime determines whether or not it will be successful in reorganizing its systems and enjoy renewal, or face a steady deterioration and devolution of its infrastructure and eventual death and decomposition of the society. The oil-based civilization, the most successful energy regime in all of human history, is just a few short years away from the turning point. The paths of three defining forces are quickly converging, forcing society to make decisions about what steps to take to ensure the future. The interplay between the imminent peak of global oil production, the increasing concentration of the remaining oil reserves in the Middle East (the most politically and socially unstable region of the world), and the steady heating up of the world's atmosphere from the spent energy—or entropy—accumulated over the course of the Industrial Age make for a volatile and dangerous world game, one whose outcome at this stage remains very much in doubt.

Fossil-fuel civilization is under siege. How well we respond to the triple threat at our front gates depends, in part, on how vulnerable our existing infrastructure is to attack, disruption, and disrepair. On this score, the prospects are poor. The complex, centralized infrastructure we have created to manage a high-energy, fossil-fuel economy—once our greatest asset—is fast becoming our biggest liability. We are increasingly vulnerable to threats and disruptions from without and within, making this moment in post-industrial history the most precarious in memory.

7

VULNERABILITIES ALONG THE SEAMS

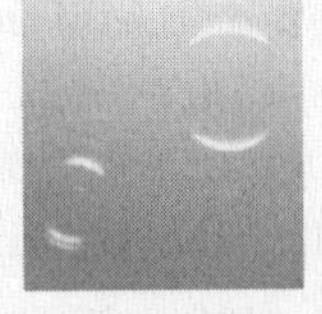

On September 11, 2001, nineteen Arab terrorists, armed only with box cutters, commandeered four U.S. commercial jetliners in flight. Two of the planes struck the World Trade Center towers, toppling the fifth- and sixth-tallest office buildings in the world. The third plane struck the U.S. Pentagon in Washington, and the fourth plane went down in Pennsylvania as passengers fought furiously to regain control of the aircraft.

It was the worst terrorist attack on U.S. soil in its 225-year history. More than three thousand people on the planes and in the buildings were killed. We later learned that fifteen of the hijackers were Saudi Arabians connected to a radical Islamic paramilitary sect called the al Qaeda network, led by Osama bin Laden, the Saudi Arabian terrorist responsible for earlier attacks on U.S. embassies in Africa and the U.S. warship *Cole* in Yemen, and for the 1993 bombing of the World Trade Center.

Bin Laden had fought alongside other Arab "freedom fighters" in Afghanistan in the 1980s in a holy war to rid that Muslim nation of Russian troops. Bin Laden first turned his ire against the U.S. in the 1990s during the Persian Gulf war. Iraq invaded Kuwait in August of 1990 in an attempt to annex its rich oil fields. The U.S. marshaled a global coalition to repel the attack. Five hundred thousand American soldiers were called up for duty, and, in January 1991, the American military—with the support of 160,000 additional troops from other allied nations (34 countries joined the coalition)—launched a massive air and land assault, successfully routing the Iraqi military and expelling them from Kuwait. One hundred thousand Iraqi soldiers died in the conflict, and 148 American soldiers also died in the war.[1]

Critical to the war effort was an agreement reached between the U.S. government and the Saudi royal family to allow American jet fighters to use Saudi airfields and airspace. The American military presence continued after the war was concluded. As mentioned, bin Laden and other radical Islamists were enraged that the Saudi government had allowed the "infidels" to desecrate the holy land of Islam by maintaining a presence there. Bin Laden vowed a jihad to drive the Americans out of his country. His subsequent terrorist attacks on American installations were aimed at achieving that objective.

On September 11th, he brought his jihad to U.S. shores for a second time, in a suicide attack mission against its civilian population and military personnel. Sadly, he succeeded beyond even his own wildest expectations. A shocked nation responded. President George W. Bush rallied the world community for a sustained war against terrorism. The U.S. government launched air strikes against the radical Islamic regime, the Taliban, which controlled Afghanistan and which had given refuge to bin Laden and his terrorist fighters. The air strikes were followed by a ground assault led by Afghan freedom fighters—the Northern Alliance—who had been engaged in a protracted struggle for years to free their country from Taliban rule. In December 2001, the Taliban regime fell, and thousands of Taliban troops and al Qaeda fighters were captured. Many others in the al Qaeda network fled into Pakistan. Meanwhile, the al Qaeda terrorists cells, thought to be active in fifty countries, were brought under surveillance. Terrorists were arrested, and assets of organizations suspected of laundering funds to the terrorists were frozen by the U.S. and other governments. Still, the network continues, as do other extremist Muslim groups, including Hamas, Islamic Jihad, and Hezbollah.

The September 11th attacks were part of a continuum of events that began a decade earlier in the Persian Gulf war and whose roots lie even further back in the early colonial years, when Western governments and oil companies, in consort with Arab strongmen, established their presence in the Middle East for the purpose of securing cheap crude oil for the industrial machine.

Since World War II, access to Middle East oil reserves has been a critical dimension of U.S. foreign policy. British, French, German, Dutch, and other European powers were engaged in the region even earlier. For much of the Cold War, the U.S. and the former Soviet Union vied to secure a dominant foothold in the oil-rich Gulf.

The military cost alone of protecting U.S. oil interests in the Middle East has been enormous. For more than half a century, the U.S. military has had to maintain naval fleets, airfields, and other military installations and personnel in and around the Gulf to guarantee the safe passage of oil from that region to the U.S. and other consuming nations. The Persian Gulf war marked a turning point of sorts. For the first time, the U.S. began spending more money to protect its vital interests in that part of the world than it received in terms of the value of oil imported. As in the Roman Empire in its later stages, the military costs of maintaining the energy flow-through was becoming more expensive than the net value of the energy being secured. Marginal returns had set in. The figures were staggering.

The U.S. government is estimated to have spent between $61 billion and $71 billion in additional funds in deployment, construction, and operations in the Gulf War—and those are funds above and beyond the monies already spent on existing military equipment used in the conflict. Fifty-four billion dollars of the funds were offset by contributions from allies.[2] Since that time, U.S. military presence in the Persian Gulf has continued, with 9,000 troops on land (mostly in

Saudi Arabia) and 15,000 at sea. Hundreds of U.S. warplanes and dozens of ships patrol the region, and as many as 50,000 U.S. troops take part in joint war exercises there each year.[3] It is estimated that the U.S. government currently spends between $30 billion and $60 billion per year defending its oil interests in the Middle East. Others put the figure between $20 billion and $30 billion.[4]

When the economic costs of the September 11th attacks are heaped on top of the existing costs in maintaining America's military presence in the Middle East and the new costs incurred in the Afghan war, the numbers begin to look even more foreboding. Economists estimate a $30-billion price tag to pay for the physical damages and cleanup costs of the September 11th attacks. But that's just the beginning. New York City is projected to lose an additional $4 billion to $5 billion in retail sales in the twelve months following the attacks and between $7 billion and $13 billion in tourist revenue by the end of 2003.[5] New York City government officials are also worried about the long-range prospects for the city. New York is already the most expensive metropolitan area in the country in relative business costs, and there is concern that the increase in costs and inconvenience to maintain beefed-up security for the city—especially in the financial district, which is considered ground zero in the terrorists' war to undermine American economic presence in the Middle East and around the world—could dissuade businesses from locating in New York as well as convince existing firms to leave the city altogether.[6] Many financial companies with offices in and around the World Trade Center experienced heavy losses in personnel and revenue in the attack and its aftermath and are beginning to ask whether it is worth maintaining their operations in what is now regarded as a high-risk zone.

The terrorist attack has also had a profound effect on the lives of New Yorkers. Beside the psychological stress of living in a small, densely populated island with only a few escape routes over bridges and through tunnels, there is also the deteriorating economic situation, which many believe is likely to linger on for years. More than 109,000 New Yorkers lost their jobs at airlines, travel agencies, hotels, restaurants, and car rental agencies in the two months following the attack. These are jobs above and beyond what would have been lost if there were only a recession to contend with.[7]

The economic fallout has spread far beyond New York, affecting the entire global economy. The forecasting group Macroeconomic Advisers estimates that the attack resulted in more than $13 billion in losses in private and government capital. Boeing, for example, let go more than 30,000 workers in anticipation of a precipitous drop in new orders for commercial jetliners. Macroeconomic Advisers calculates that the attack lowered economic activity in the third quarter of 2001 by a whopping $24 billion.[8]

The Milken Institute, an economic research group in Santa Monica, California, reports that, by the end of 2002, more than 1.8 million jobs will have been lost in the United States because of the terrorist attacks. Outside New York City, the cities where losses will likely be greatest, according to the Milken study, are Los Angeles, Chicago, and Las Vegas.[9]

Nowhere has the ripple effect of the September 11th attacks been more apparent than in the losses in the travel and tourism industry, one of the largest industries in the world. In the year 2000, the industry generated $4.5 trillion in revenue and accounted for 11 percent of the GDP of the world and 8.2 percent of total world employment, or 207,062,000 jobs.[10]

The World Travel and Tourism Council estimates a 10 to 20 percent decline in travel and tourism in the U.S. in the twelve months following the attack, and less for other countries.[11] A 10 percent decline in revenue in the U.S. would amount to $48 billion in losses, while a 20 percent decline would top $96 billion.[12] The council says that a likely decline of 10 percent in worldwide travel and tourism will mean a 1.7 percent fall in worldwide GDP and a loss of 8.8 million jobs.[13]

The attacks also cost American taxpayers. The war in Afghanistan cost about $1 billion per month, and the U.S. government was expected to earmark $28 billion in funds for military-related activity in the region in the following year.[14] Moreover, the U.S. pledged $300 million in aid to the new government of Afghanistan to help in reconstruction efforts in that war-torn country.[15] In addition to the increased military costs in pursuing the war on terrorism abroad in Afghanistan and elsewhere, taxpayers gave the families of victims of the attacks in New York and Washington payments averaging $1.65 million, with an overall estimated price tag of $6 billion. The federal government also committed a total of $11.2 billion in federal aid to assist New York City in its recovery efforts, as well as $15 billion in funds to bail out the airline industry.[16]

Finally, the new federal office of Homeland Security, whose mission "will be to develop and coordinate the implementation of a comprehensive national strategy to secure the United States from terrorist threats or attacks," was given more that $19 billion in federal funds for its efforts in 2002 and was expected to receive $37 billion more for programs in 2003.[17] An additional $3.5 billion in emergency money was allocated to the military—beyond funds already included in the military budget—to assist in the military campaign against Osama bin Laden and the al Qaeda network.[18] The National Governors Association estimated that another $4 billion in funds would be needed at the state level to provide for increased security costs in the aftermath of the attack.[19]

Bear in mind that all of this cost in lost lives and losses to the U.S. and world economy and to U.S. taxpayers resulted from the suicide mission of just nineteen Islamic radicals armed with box cutters and motivated by a deep hatred of the role of the United States in the world, and especially its military presence in Saudi Arabia and the Middle East. The lesson here is that in a dense, highly complex urban society like the metropolitan area of New York City, the ability of a handful of determined terrorists to wreak havoc and cause death and mass destruction is high.

Bioterrorism

The vulnerability of human populations clustered close together in high-occupancy social environments came home two weeks after the attacks on the World Trade Center and the Pentagon when a deadly strain of anthrax began showing up in letters sent to newsrooms of national media and shortly thereafter to congressional offices in the nation's capital. The weapon-grade anthrax—in the form of a white powder—was sent through the U.S. mail to its victims. Eighteen Americans were exposed to the anthrax spores, and five subsequently died from inhaling the deadly pathogen. Terrorist notes were found with letters praising Allah and condemning the U.S. government for its policies.

Just days earlier, the FBI had reported that several of the hijackers responsible for the September 11th attacks had made a number of visits in the preceding weeks to a facility in Florida that houses crop-dusting planes. According to the proprietors, the hijackers asked questions about load capacity, range, and operability of the specialized planes. The FBI subsequently ordered all 3,500 of the nation's privately owned crop dusters grounded pending further investigation. Meanwhile, universities, including the University of Michigan, Penn State, Clemson, and Alabama, barred aircraft from flying over their stadiums during football games, for fear of biowarfare attacks.[20] In Washington, policymakers scurried to catch up to mounting public anxiety by allocating funds to stockpile antibiotics and vaccines and upgrading emergency procedures at the nation's hospitals and clinics.

To date, federal authorities are unsure of whether the spread of anthrax in the mail is connected with the al Qaeda terrorist network. What they do know is that Osama bin Laden's network of terrorists has actively pursued information on biological agents, including how to procure them, produce them in large volume, upgrade their lethal properties using state-of-the-art biotech databases, and find ways to distribute them over wide geographic areas.

The anthrax terrorist assault highlights how unprepared America and the world are for such attacks, and especially how vulnerable concentrated urban populations are to this deadly form of warfare. Even as the fear of anthrax gripped an already frightened nation, politicians, military experts, and the media skirted a far more disturbing reality that lies at the very heart of the new fears about bioterrorism. The new genomic information being discovered and used for commercial genetic engineering in the fields of agriculture, animal husbandry, and medicine is potentially convertible to the development of a wide range of novel pathogens that can attack plant, animal, and human populations.

Moreover, unlike nuclear bombs, the materials and tools required to create biological warfare agents are easily accessible and cheap, which is why this kind of weapon is often referred to as "the poor man's nuclear bomb." A state-of-the-art biological laboratory could be built and made operational with as little as $10,000 worth of off-the-shelf equipment and could be housed in a room as small as 15 square feet. All you really need is a beer fermenter, a protein-based culture, plastic clothing, and a gas mask. Equally frightening, thousands of molecular biologists and graduate-school students in laboratories around the world are knowledgeable enough in the rudimentary uses of recombinant DNA and cloning technology to design and mass-produce such weapons.

Biological warfare (BW) involves the use of living organisms for military purposes. Biological weapons can be viral, bacterial, fungal, rickettsial, or protozoan. Biological agents can mutate, reproduce, multiply, and spread over a large geographic terrain by wind, water, insect, animal, and human transmission. Once released, many biological pathogens are capable of developing niches and maintaining themselves in the environment indefinitely.

Biological weapons have never been widely used, because of the danger and expense involved in processing and stockpiling large volumes of toxic materials and the difficulty in targeting the dissemination of biological agents. Advances in genetic engineering technologies over the past decade, however, have made biological warfare viable for the first time.

Recombinant DNA "designer weapons" can be created in many ways. The new technologies can be used to program genes into infectious microorganisms to increase those microorganisms' antibiotic resistance, virulence, and environmental stability. It is possible to insert into organisms genes that affect regulatory functions that control mood, behavior, and body temperature. Scientists say that they may be able to clone selective toxins to eliminate specific racial or ethnic groups whose genotypic makeup predisposes them to certain disease patterns. If the intent is to cripple the economy of a country, genetic engineering can also be used to destroy specific strains or species of agricultural plants or domestic animals.

The new genetic engineering technologies provide a versatile form of weaponry that can be used for a wide variety of military purposes, ranging from terrorism and counterinsurgency operations to large-scale warfare aimed at entire populations.

Professional military observers are not sanguine about the prospect of keeping the genetics revolution out of the hands of the terrorists. As a tool of mass destruction, genetic weaponry rivals nuclear weaponry, and it can be developed at a fraction of the cost. These two factors alone make genetic technology the ideal weapon of the future.

The revelation that Iraq had stockpiled massive amounts of germ-warfare agents and was preparing to use them during the Persian Gulf war renewed Pentagon interest in defensive research to counter the prospect of an escalating biological arms race. Saddam Hussein's government had developed what it called the "great equalizer," an arsenal of twenty-five missile warheads carrying more than 11,000 pounds of biological agents, including deadly botulism poison and anthrax germs. An additional 33,000 pounds of germ agents were placed in bombs to be dropped from military aircraft.[21] Had the germ-warfare agents been deployed, the results would have been as horrible as those visited on Hiroshima and Nagasaki with the dropping of the atomic bombs in 1945.[22] To get a sense of the potential damage that could have been inflicted, compare the Iraqi arsenal with a study conducted by the U.S. government in 1993 that found that the release of just 220 pounds of anthrax spores from an airplane over Washington, D.C., could kill as many as three million people.[23]

Iraq is not alone in its interest in developing a new generation of biological weapons. In a 1995 study, the Central Intelligence Agency (CIA) reported that seventeen countries were suspected of researching and stockpiling germ-warfare agents. The nations were Iraq, Iran, Libya, Syria, North Korea, Taiwan, Israel, Egypt, Vietnam, Laos, Cuba, Bulgaria, India, South Korea, South Africa, China, and Russia.[24] It is likely that terrorists, working with rogue countries, will turn to the new genetic weapons to spread fear and chaos as they seek to have their demands met by society.

The new head of Homeland Security in the U.S., Governor Tom Ridge, has vowed to make bioterrorism the highest priority in his efforts to safeguard the American people. His task will be difficult if not impossible. Even with all the safeguards in place, there is little that can be done to prevent bioterrorists from breaking through the security net and unleashing deadly pathogens that could kill thousands, even millions, of people. Like the Arab hijackers who, armed only with box cutters, succeeded in bringing the mightiest city in the world to its knees, at least for several days, bioterrorists with deadly anthrax or other pathogens and a modicum of knowledge of how to employ them could do similar harm, further weakening American security and the world economy. The specter of, for example, coordinated smallpox attacks unleashed simultaneously by Islamic militants in different parts of the world, using new, genetically engineered strains for which there is no antidote, is almost beyond comprehension. Millions of people could die from such attacks, and the economic and social repercussions would extend far into the future.

Even if Osama bin Laden and the leadership of the al Qaeda network are killed or caught and their network routed, few believe that will be the end of the story. With more and more young Muslim youth being recruited into terrorist activity around the world, most observers expect the level of violence to only spread and deepen in intensity, requiring additional countermeasures by the U.S. and other governments to defend their vital oil interests in the Middle East and protect homeland security. The process will likely accelerate as the world moves closer to the peak of the global oil-production curve sometime in the next ten years. All the while, the escalating violence and growing military costs of maintaining secure passage for Middle East oil will continue to result in diminishing returns for the U.S. and the world's other leading industrial nations.

The Soft Spots

The escalating terrorism on U.S. soil is making Americans aware, for the first time, of how vulnerable the country is to threats to its infrastructure and population. The White House, Congress, and the Pentagon are now turning their attention to the many other existing soft spots that pose serious risks to American security. What they are finding is that our fossil-fuel infrastructure is vulnerable to disruption and collapse at a number of critical points along its seams. If energy throughput were to be seriously impeded at any of these points for an extended period of time, it could cripple the overall system.

Three developments make the present era so much more vulnerable than previous eras in history. First, there is the qualitative leap in food production made possible by the application of fossil-fuel–based fertilizers and pesticides in agricultural production and the accompanying replacement of oil-powered machines for human labor on the farm, which freed up millions of people to migrate from rural to urban areas. Second, the 20th century has witnessed the greatest surge in human population in history, thanks in part to the massive food surpluses of modern agricultural production. Third, fossil-fuel energy has allowed humanity to build a high-energy transportation system and electrical grid that connect millions of people in a tightly organized and interdependent social matrix, whose functioning is essential to every person's survival. The critical points, then, in our contemporary social architecture, are the petrochemical-based agricultural system; dense urban life; an oil-fueled transportation mode that ferries passengers and freight quickly between rural, urban, and suburban communities and across continents and oceans; and the electricity grid, the "central nervous system" that provides power, light and heat, and a telecommunications network to coordinate the workings of all the subsystems that make up the larger organism. Every one of these vital points that constitute the superorganism that we have come to think of as the industrial and information-age economy exists only because of the continuous throughput of oil and, to a lesser extent, natural gas and coal.

Whenever concerns are raised about running short of cheap crude oil, most people immediately worry about not having enough gasoline for their automobiles. The prospect that the nation's auto, truck, bus, and airplane fleets might be slowed or even brought to a halt by a shortage of fuel is frightening enough and would certainly have a devastating effect on the economy and society. But what is often overlooked is that oil and natural gas are also critical to maintaining both our food production and the electrical power system that gives us power, light, and heat. Should these systems fail us, the entire social organism would break down.

Growing Food in Oil

Agriculture provides the core energy to maintain the whole of industrial society in a state far from equilibrium. Modern agriculture is dependent almost exclusively on oil. Were oil to become more scarce, expensive, and unavailable, every other aspect of contemporary life would contract.

Our modern food production makes possible all of the other economies that ride atop it. Large food surpluses and the freeing up of farm labor made possible the 20th-century manufacturing revolution and, later, the service and information economies. And even though agricultural output has become a commodity over the past half century and food prices have remained relatively cheap and stable during that time, all of that could change overnight once global oil production peaks and sends oil prices dramatically higher.

Currently, 4 percent of all the energy consumed in the United States goes to growing food. Another 10 to 13 percent of America's total energy consumption is used to transport, process, package, and deliver that food to American supermarkets. More than 17 percent of all the energy used in the United States, then, goes to putting food on our tables.[25]

Many anthropologists consider the great expansion of agricultural yield to be the singular achievement of the modern era. That accomplishment was attained by the replacement of human labor with mechanical labor—run on oil—and by the increasing use of petrochemical-based fertilizers and pesticides to enhance yields.

The first gasoline-powered tractor was built by John Froehlich in 1892. By 1910, 25,000 tractors were being used on U.S. farms. When Henry Ford introduced the Fordson, a cheap reliable model, tractor sales climbed.[26] On the eve of World War II, 1.6 million tractors were being used, and by the 1960s the number had risen to 4.7 million.[27] Use of motor trucks also increased. In 1915, there were only 25,000 trucks on farms. By 1980, more than 3.5 million trucks were being used.[28] The gasoline engine, harnessed to tractors, trucks, and harvesting machines, has replaced human labor, horses, mules, and oxen in less than a century as the primary source of power on the farm.

The mechanical revolution in agriculture in the first half of the 20th century was followed by a chemical revolution in the second half of the century. From 1950 to 1989, the use of inorganic nitrogen fertilizers made from petrochemicals rose from 14 million tons to 143 million tons.[29] The use of pesticides increased from 200,000 pounds in 1950 to more than 6.4 billion pounds in 1986.[30]

The mechanization of agriculture and the use of petrochemical-based nitrogen fertilizers and pesticides, along with advances in breeding and the introduction of high-yield monoculture crops, increased food production dramatically over the course of the 20th century while reducing the amount of human labor needed on the farm. In 1850, 60 percent of the U.S. labor force was employed in farming. Today, less than 2.7 percent of American workers are directly engaged in agriculture.[31] At the same time, agricultural yields have skyrocketed. In 1850, a single farm produced enough food to feed four people. By 1982, a single farmer produced enough food to feed seventy-eight people.[32] Agricultural productivity increased by 25 percent in the 1940s, 20 percent in the 1950s, 17 percent in the 1960s, and more than 28 percent in the 1980s.[33]

The gains in yield and productivity, however, have been at the expense of more and more oil used in the process. From a thermodynamic perspective, modern agriculture has been the least productive form of agriculture in history. That is, it has used far more energy inputs per unit of energy output than in any previous period. A peasant farmer usually produces about ten calories of output for each calorie expended. Although an Iowa farmer, applying the most advanced technology available, can produce 6,000 calories for every calorie of human labor expended, the achievement becomes less significant when we calculate the amount of energy used to produce net energy returns. To produce one can of corn containing 270 calories, the farmer uses up to 2,790 calories to power the machinery and provide the synthetic fertilizers and pesticides. So, for every calorie of energy actually produced, the high-tech American farm ends up using ten calories of energy in the process.[34]

The increased energy flow-through has also resulted in more entropy in the surrounding environment. As the native soil base has depleted and eroded because of intensified farming practices, more synthetic fertilizers have had to be applied to sustain the yields. Nitrate pollution from fertilizer runoff now accounts for half of our water pollution and two-thirds of our solid-waste pollution.[35] Monoculture practices—growing a single crop over extended acreage—have created economies of scale and increased productivity and profits but have required the use of greater amounts of pesticides. More-traditional farming relied on the planting of diversified crops that attracted a range of insects, some of which are natural enemies of insect pests. Eliminating crop diversity in favor of monoculture crops left the fields without the beneficial insects, and crops became more vulnerable to insect pests, requiring a steady rise in the use of pesticides. Much of the sprayed pesticide runs off into the groundwater and becomes a major source of water pollution in every agricultural region of the world.

The pesticides also destroy the remaining soil. The soil contains millions of microscopic bacteria, fungi, algae, and protozoa, as well as worms and anthropods. These organisms maintain the fertility and structure of the soil. Pesticides destroy these organisms and their complex habitats, hastening the process of soil depletion and erosion. American farms lose more than four billion tons of topsoil annually, much of it because of the high-tech farming practices introduced over the past half century.[36] By the 1970s, the U.S. had lost more than one-third of its agricultural topsoil.[37] The depletion and erosion, in turn, have required the use of ever-increasing amounts of petrochemical fertilizers to maintain agricultural output. Marginal returns have set in. More and more energy inputs are required to produce smaller gains in net energy yield. Between 1945 and 1970, for example, American corn farmers increased their energy inputs by 400 percent but only increased yields by 138 percent. In the first two decades of intensive petrochemical-based farming following World War II, overall energy inputs in agriculture rose by 70 percent, but food production only increased by 30 percent.[38] Moreover, high-energy agriculture is now a major contributor to global warming. Greater mechanization has meant burning more gasoline, increasing CO_2 emissions, while greater reliance on petrochemical fertilizers has increased the release of nitrous oxide, a potent global-warming gas.

The mounting costs of maintaining a technologically complex, high-energy agriculture have put most small family farms out of business. The new farming requires economies of scale and large infusions of capital investment. Fewer than 32,000 large-scale farms in the United States now account for more than 38 percent of farm-product sales.[39]

The chemical revolution in the U.S. and Europe spread into the developing world in the 1960s—the so-called "Green Revolution." Countries in Asia were especially hard-hit by a fast-growing population in a region of the world where every available piece of land had long since been put into agricultural production. To meet the demands of a growing population, scientists created new super-strains of wheat and rice that greatly increased yield per acre. These high-yielding varieties doubled production in less than a decade in places like India and Pakistan, but they required vast inputs of petrochemical fertilizers and chemical pesticides. The monoculture crops, many of which were grown for export on world markets rather than for local consumption, also did best under economies of scale that increased the capital costs and had the effect of edging more subsistence farmers off the land to make room for large farm holdings. Throughout the Third World, impoverished peasants have been forced to migrate to large cities, where many live off the public dole or remain destitute on the streets.

We have built a worldwide agricultural infrastructure run by fossil fuels, and the short-term increase in yield has made it possible to greatly expand both the total human population as well as the number of people living in urban areas. Now, with global oil production likely to peak sometime within the next decade or so and make oil increasingly expensive, how do we sustain enough agricultural yield to maintain a growing human population into the 21st and 22nd centuries? Even though we are still awash in relatively cheap oil and high yields, nearly one billion people are undernourished.[40] Imagine the prospects when global oil production peaks, sending oil prices irreversibly higher.

The life-science companies tout the new genetically engineered food crops as a solution. But these crops, too, require large inputs of energy, especially in the form of petrochemical fertilizers. Scientists have thus far been unsuccessful at creating biotech crops that get their own nitrogen from the air rather than from the soil. And studies show contradictory results in terms of yield performance. The future prospects for agriculture look even bleaker when we consider the fact that 11 percent of the land surface on the planet is already used to produce food, which leaves little decent land left for agriculture.[41] In desperation, human beings have begun to cut down large swaths of tropical rain forest in the Amazon and elsewhere to make room for agricultural production. The destruction of the rain forest eliminates precious habitat for many of the Earth's remaining species of plant and animal life. The soil base itself is too thin to support food production for more than a few years. The result is spreading erosion and barren land, no longer fit for human, animal, or plant habitation.

To make matters worse, one-third of the world's agricultural land has been converted from growing food grains for human consumption to growing feed grain for cattle and other livestock. Cattle production is now the most energy-consuming agricultural activity in the world. It takes the equivalent of a gallon of gasoline to produce a pound of grain-fed beef in the U.S.[42] To sustain the yearly beef requirements of an average family of four people requires the consumption of more than 260 gallons of fossil fuel. When that fuel is burned, it releases 2.5 tons of additional CO_2 into the atmosphere—as much CO_2 as the average car emits in six months of normal operation.[43]

If well-off consumers in the West and elsewhere were willing to forego a heavy meat diet and eat farther down on the food chain with a largely vegetarian diet, precious farmland could be freed up to grow food for millions of people. But a change in food habits in the West would have to be accompanied by land reform and other structural changes in the Southern Hemisphere so that the poor could regain access to the land to grow food grain for their families and communities.

Government leaders, economists, and consumers anguish over

the prospect of running short on gasoline to power automobiles. Of far greater concern, however, is the prospect that the price of growing food itself might become so expensive, as we move to the backside of the crude-oil bell curve, that hundreds of millions of people, perhaps even billions, will not be able to afford to buy sufficient food to feed themselves and their families. Although alternative sources of energy for fueling automobiles and trucks already exist (and will be discussed in detail in chapter 8), no such substitutes exist for petrochemical fertilizers. That means that for every gallon of gasoline we use now to power our vehicles, we are left with one less gallon to grow the world's food later. The tradeoff is troubling. Consider an automobile that gets thirty miles per gallon of fuel burned. Every six miles traveled in that vehicle burns enough gasoline to produce one loaf of bread.[44] In the future, we may have to make uneasy tradeoffs between personal mobility and eating. Surprisingly, in all of the discussion surrounding oil and the future of the fossil-fuel age, not a single world leader has been willing to address this prospect publicly.

The fossil-fuel energy surge has led to an explosion of human population over the past 150 years. It took from the very beginning of human history until 1825 for the human population to reach one billion. With the onset of the coal revolution and a steep rise in energy throughput, the human population doubled to two billion in less than one century. Another billion people were added between 1925 and 1960 with the birth of the oil age. World population rose to four billion between 1960 and 1975 and to five billion twelve years later, in 1987.[45]

The rapid growth in human population resulting from more and better nutrition, as well as sanitation, created a positive feedback. More people require more complex social organization, which means harnessing even greater stores of available energy to maintain institutional arrangements. The nation-state is a prime example of the energy process at work. Rail travel and the telegraph allowed commerce and trade to expand across wider geographic areas. Expanded commerce and trade, in turn, gave rise to a new political unit, the nation-state—a governing arrangement expansive enough to secure far-flung resources, mobilize diverse labor pools, and coordinate mass consumer markets. Social historian Charles Tilly estimates that Europe, at the beginning of the 16th century, was made up of more than 500 small autonomous governments—city-states, principalities, and kingdoms. By 1975, the number of governing units had been reduced to only thirty-five.[46] In the 21st century, with faster land and air transport and speed-of-light communications shortening distances and compressing time even further, Europe is quickly becoming a single governing entity. The European Union adopted a single currency in 2002 and is likely to expand over the course of the next few years from its fifteen member nations to twenty-seven countries, making them all part of a continent-wide governing structure stretching from the Atlantic coastline to the former Soviet Union.

Even more impressive is the spectacular growth of high-energy throughput mega-cities during the fossil-fuel age. Cities before the current era were much smaller in scale. Ancient Babylon, at its height, had only 100,000 inhabitants. The population of Athens was less than 50,000. Around 1820, London became the first city of the fossil-fuel era to reach a million inhabitants. One hundred years later, only eleven cities in the world exceeded one million. By 1950, however, seventy-five cities numbered over one million inhabitants, and by 1976, 191 urban areas contained more than one million residents apiece.[47] Today, cities all over the world have populations exceeding one million people. Nineteen cities, including Tokyo, Mexico City, Bombay, São Paulo, Shanghai, and New York are mega-cities with populations ranging between ten and twenty-five million.[48]

Just two centuries ago, most human beings lived in rural areas and small villages and towns. Today, over half the human population on Earth lives in densely populated urban areas. These cities and their suburbs exist by sucking vast sums of available energy from their surrounding environments. They require a steady flow of food, energy, water, and mineral resources to maintain and sustain them. Just to maintain an urban existence, an average city of one million people requires more than four million pounds of food, 625,000 tons of freshwater, and 9,500 tons of fuel every twenty-four hours, most of which has to be transported over long distances. City infrastructures are also heavy energy users. The Sears Tower in Chicago, for example, consumes more energy in twenty-four hours than does a city of 150,000 people.[49]

Cities also exist because the internal combustion engine on wheels connects farm and town, suburb and city center, in a seamless transportation network. The shipping of agricultural products thousands of miles from producing regions to urban areas would not have been possible before the advent of the railroad, truck transport, and modern refrigeration storage processes—all powered by burning vast amounts of oil.

The point that needs to be emphasized is that supporting more than half the human race in urban environments would not be conceivable were it not for the increase in agricultural yield and productivity made possible by the use of oil to power farm machinery, fertilize land, fend off agricultural pests, and transport products to faraway metropolitan areas. Cities, then, are precarious arrangements perched atop a fragile and vulnerable agricultural foundation. They will continue to exist only as long as agricultural production can sustain them.

When the Electricity Goes Off

It is understandable that we would be unmindful of the critical role that oil plays in feeding our families, because the process of growing food is so removed in time and place from our urban lives. The same holds true for the electricity that we have come to rely on to maintain our daily routines. The electrical grid is the central nervous system that coordinates a densely populated urban existence. Without electrical power, urban life would cease to exist, the information age would become a faded memory, and industrial production would grind to a halt. The fastest way to ensure the collapse of the modern era would be to pull the plug and turn off the flow of electricity. Light, heat, and power would all stop. Civilization as we know it would come to an end.

It is hard to imagine what life would be like without electricity, although it has only been utilized as a source of energy for less than a century. Most of our great-grandparents were born into a world without electricity. Today, we take electricity for granted. That is because, like our food, it is abundantly available. We rarely think about where it comes from or how it gets to us. It is a kind of stealth force, tucked away inside wires overhead, buried in the ground, or hidden inside our walls. Colorless and odorless, it is an invisible but indispensable presence in our lives.

How many of us have ever been to a power plant and watched how our electricity is actually generated? Most of us tend to think of electricity as a primary source of energy, not stopping to realize that most of the electricity we use is generated by burning coal, oil, or natural gas.

How likely is it that the electricity might go off, not just for a brief moment but for extended periods of time? Unfortunately, the nation's electrical power grid is increasingly vulnerable to disruption, both by terrorists and energy shortages. Even before the September attacks, government officials worried that American power plants, transmission lines, and the telecommunications infrastructure could be targets for terrorists. In 1997, the President's Commission on Critical Infrastructure Protection issued a warning that cyber-terrorists' next target might be the computer programs at the power-switching centers that move electricity around the country. Disrupting the electrical grid could wreak havoc on the nation's economic and social infrastructures. Richard A. Clarke, who heads the cyber-terrorism efforts of the Bush administration, warns of an "Electronic Pearl Harbor." A report issued by the Institute for Security Technology Studies shortly after the September 11th attacks cautioned that the American military reprisal against the al Qaeda network and the Taliban in Afghanistan could be met by a counter-retaliatory strike by terrorists against the American electronic infrastructure. Jeffrey A. Hunker, dean of the Heinz School of Public Policy and Management at Carnegie Mellon University and formerly senior director for Protection of Critical Infrastructure at the National Security Council, believes that the nation is "sitting on a cyber time bomb."[50]

The U.S. government has recently established a national Infrastructure Protection Center, a collaborative effort between law-enforcement agencies at the local, state, and federal levels to protect against computer crime aimed at disrupting America's electronic infrastructure, but, privately, many experts in the field of electronic security are not confident about the possibility of defending the nation's electricity grid. The President's Commission's own report noted that, by the year 2001, more than 19 million people around the world would have the requisite computer skills to create minor disruptions to the nation's electric power structure, and 1.3 million people would have the kind of sophisticated knowledge of how the nation's electrical power grid and telecommunications network operate to create significant damage.[51]

America's electrical power grid has experienced serious power failures on a number of occasions over the past thirty-seven years, each one creating panic and a taste of what might happen if blackouts were to become more frequent and lengthy in duration. The first major blackout occurred on November 9, 1965. A single malfunctioning relay in Canada resulted in a cascading power failure that quickly enveloped most of the northeastern United States, plunging the region into darkness. More than thirty million Americans lost all electric power for more than twelve hours. The states of New York, Connecticut, Massachusetts, Vermont, and Maine were all affected. In New York City, people were stuck in elevators and traffic was snarled to near-gridlock as traffic lights went out throughout the metropolitan area. Power, light, and telephone services were knocked out everywhere. An eerie silence hung over New York City and the rest of the Northeast. In a moment of time, millions of people, so accustomed to a world mediated by electricity, suddenly found themselves vulnerable and without recourse. The many conveniences they had come to rely on to sustain their lives went dead. A kind of village life returned to the hardened streets of New York for several hours as residents descended onto the sidewalks seeking information and comfort from strangers-turned-neighbors. This first time, the police reported an actual drop in crime in the city of New York. Nine months later, New York health authorities announced, in a somewhat humorous vein, a surge in new births. Apparently, the loss of television reception turned more than a few New Yorkers to more traditional forms of entertainment.[52]

Still, the public was enraged by the outage. Investigations were launched, recommendations were made, and changes were implemented to guarantee that a blackout would never happen again. But it did happen again, despite all the public assurances. On the evening of July 13, 1977, lightning struck a tower in Westchester County, north of New York City, and short-circuited two high-voltage power lines, triggering a cascade of events that shut down the electrical grid in and around New York City. The blackout affected nine million people. Electricity remained out of service for more than fifteen hours. Unlike the 1965 blackout, which occurred on a cool November evening, this second occurrence struck at 9:34 P.M. on a hot, humid night, shutting off air conditioning and sending human emotions to the boiling point. In poorer neighborhoods of the city, mobs of people went on a rampage, burning buildings and looting stores. More than 4,000 people were arrested in the blackout, and seventy-eight policemen were injured in the melee. The chairman of the board of Consolidated Edison, the electric utility company that managed the power grid for New York City and its suburbs, called the blackout "an act of God." Surveying the damage done by looters in the Bushwick neighborhood of Brooklyn, a Roman Catholic priest gave the event a different spin, proclaiming, "We are without God now."[53]

The western coast of the United States has experienced similar power failures and massive blackouts. The first major outage occurred at 3:45 P.M. on August 10, 1996, and extended from Oregon to the Mexican border. Nine states in all were affected by the blackout, which occurred on one of the hottest days of the year with temperatures reaching 113°F. Five million California residents lost all power. The blackout was traced to power lines in Oregon that were sagging due to the excessive summer heat. The power lines touched some overgrown trees, triggering the outage. The event created a chain reaction that rippled across the West Coast, shutting off power in region after region.[54]

Rolling blackouts hit California again in March 2001, affecting 800,000 people from Oregon to Orange County in Southern California. This time, the reason was that not enough electricity was being generated by power plants to handle a spike in energy demands caused by unusually warm spring weather.[55] Unlike the earlier blackouts that hit the Northeast, the ones on the West Coast shut down computer systems in businesses throughout the region, and this had a serious effect on commerce. In the interim quarter century, the nation had become increasingly dependent on computers and the Internet and intranets for information exchange, data storage, commercial transactions, banking and credit flows, and a host of other basic and vital services. The ripple effect of a blackout of even a few minutes in duration is now far more serious than it was in the past, when computers and software connections played a smaller role in day-to-day life.

It's worth noting that, while blackouts seriously disrupt the nation's information superhighway, the surge in computer use has also had the effect of putting additional strains on the power grid in the U.S. and other countries, making electricity shortages more likely in the future. Although it is true that microprocessors are becoming more efficient and able to process greater stores of information in shorter time periods using less electrical load, the overall demand for digital power is increasing even faster than the greater efficiencies that are coming on line. One pound of coal is needed to "create, package, store, and move two megabytes of data," notes cyberspace analyst Peter W. Huber.[56] The result is that the demand for horsepower to power personal computers is doubling every few years. Huber says that not enough attention is being paid to the fact that "chips are running hotter, fans are whirring faster, and the power consumption of our disk drives and screens is rising."[57]

Although early enthusiasts of the cyberspace revolution liked to say that huge amounts of energy would be saved in transportation costs and maintenance of brick and mortar by conducting more business and social life in virtual reality—all of which is true—it's only recently that we've become aware of the escalating electrical-power needs being generated by the migration to cyberspace. Huber points out that every single one of the 50 billion integrated circuits and the 200 billion microprocessors produced each year is powered by electricity. Chips run on electric power and require high-power densities—an average personal computer, says Huber, requires 1,000 watts of power. With the typical Internet user online some twelve hours each week, according to a study conducted by IntelliQuest, nearly 1,000 kilowatt-hours of electrical consumption is probably taking place yearly. Magnify this by the 50 million PCs in homes, the 150 million computers in businesses, and the 20 million people starting to use the Internet each year, just in the United States, and the magnitude of the challenge facing the nation's power grid becomes apparent. "For the old thermoelectrical power complex," says Huber, "widely thought to be in senescent decline, the implications are staggering."[58] And we are just in the early stages of the cyberspace revolution. The wireless web, says Huber, requires even more electrical power because the signals are broadcast in every direction rather than directed through a fiber cable. The wireless industry is already projecting the building of 70,000 radio base stations in the years immediately ahead and twice that number over the course of the coming decade. Each station will burn a couple of kilowatts of electricity.[59]

A Nation at Risk

In the wake of the 1970s oil embargo by OPEC producers and the subsequent worldwide energy shortages that resulted, the Defense Civil Preparedness Agency of the Pentagon commissioned a study to assess the vulnerabilities and weaknesses of America's energy system. The final report, released on November 13, 1981, by the Federal Emergency Management Agency, although little noticed at the time, provides a penetrating look into the many vulnerabilities of America's energy system. The picture painted by the report is of a highly complex, interdependent, and fragile energy system prone to wholesale disruption by terrorist threats, technology breakdowns and disrepair, natural disasters, and energy shortages.

The government report dealt with virtually every aspect of the nation's energy system, but of particular importance was the analysis of the shortcomings of the nation's electrical power grid. The problems highlighted in the study have remained, for the most part, unaddressed to this day, and they now pose a far greater risk to the economy and the country's security than ever before because of the triple threats of a global peak in production of crude oil and natural gas, escalating terrorist threats aimed at the nation's infrastructure by Middle East extremists, and rising temperatures on Earth because of the burning of fossil fuels. Energy analysts Amory and Hunter Lovins of the Rocky Mountain Institute in Snowmass, Colorado, who were commissioned to do the report, outlined some of the critical findings of the study.

For starters, the energy used to power the U.S. electrical grid is often located far from the end user. The average barrel of oil, for example, taken from domestic wells, travels between 600 and 800 miles before reaching its final destination. Many power plants are also located at great distances from consumers. Electricity travels an average of 220 miles or more from the power plant to the customer.[60] That means that both the energy and the electricity are more vulnerable to both adverse weather-related problems and sabotage.

Because of the high capital costs of building new power plants, the long lead time in getting them on line, and the even longer operational life expectancy of the plants themselves—some power plants remain on line for several decades—the system itself is relatively inflexible and cannot be easily or quickly modified to meet unexpected challenges, such as short-run interruptions in energy supplies, too much (or too little) demand, or a depletion in overall available energy reserves.[61] As mentioned earlier, most American power companies have invested heavily in a next generation of gas-fired power plants. But if some of the new studies by geologists are correct and global production of natural gas peaks shortly after the peak in global oil production around 2020, the nation's utilities might face shortages in natural-gas supplies and be forced to reduce the amount of electricity that they generate, resulting in serious disruptions to the economy and society at large.

Unlike primary forms of energy, electricity can't really be stored. It is, rather, a continuous flow. The moment it is generated at the power plant, it must travel through the transmission lines to the end user. Because there is no storage capacity, if there is a disruption at any point in the flow line, there is often no surplus energy readily available to make up for the loss.[62]

When the electrical grid fails, everything that relies on electricity within the geographic area is affected. A blackout means that everything that's plugged in stops working. Power failures are catastrophic because they are all-encompassing. No activity reliant on electricity is spared. So, even if it is more critical to maintain electricity in the emergency room of a hospital or in the air-traffic-control tower of a commercial airport than to assure that electricity continues to flow to a home refrigerator, the grid cannot discriminate. When the power stops, it does so throughout the system. That's why critical services and activities increasingly rely on emergency generators located on-site.[63]

Interruption of electricity also affects the flow of other primary sources of energy. Home furnaces burn on oil or gas but need electricity to ignite them, pump the fuel, and distribute the heat. Similarly, the neighborhood gasoline pumps run on electricity from the grid. Municipal water plants and sewage-treatment plants also operate with electricity. Oil refineries also depend on electricity for their operations, as does the equipment that pumps oil at the wellhead.[64]

Oil and gas pipelines that supply energy to refineries and to power plants are so complex that they are currently run by very sophisticated computer software programs and are managed by highly trained and elite workforces. If cyber-terrorists were able to penetrate these programs, they could shut down the system for a period of time, forcing the company to operate the pipelines manually with all of the increased risks and costs that accompany it.[65]

Today's power plants are so big, and the machine parts so specialized and expensive, that inventories are only kept for the small incidental items. If a major component of the system breaks, it often has to be special-ordered—and the manufacturing of the part can take up to a year or more. After the 1977 blackout in New York City, for example, Consolidated Edison ordered a spare for a phase-regulating transformer whose unavailability had played a part in the outage. It took more than a year to manufacture the item.[66] Moreover, much of the power-generating process is now automated and extremely sophisticated, requiring special contractors to repair damage. If there were multiple breakdowns at power plants scattered across the country, either because of terrorist attacks or a natural disaster, there would likely not be enough experienced personnel available to make repairs in a timely manner.[67]

Oil and gas pipelines feeding primary energy to refineries and power plants are among the most vulnerable parts of the energy matrix. The 1981 Pentagon report drew attention to the possible ramifications if any of the major pipelines in the U.S. were to be damaged by terrorist attacks. Now, twenty years later, Congress, the Pentagon, and the energy industry are revisiting the earlier study, worried that theoretical threats of a generation ago could now become realities.

Consider, for example, the Colonial Pipeline System that extends for 2,000 miles, whose three large pipes move oil from Texas all the way to New Jersey in twelve days. The operation is powered by eighty-four pumping stations and uses vast amounts of electricity to move the oil through the system. A study of the security risks attendant to the pipelines concluded that

> [they] were constructed and are operated with almost no regard to their vulnerability to persons who might . . . desire to interfere with this vital movement of fuel. They are exposed and all but unguarded at innumerable points, and easily accessible even where not exposed over virtually their entire routes. . . . [T]his vulnerability of the most important energy transportation systems of the nation threatens the national security. . . . Although all forms of energy movement are vulnerable to some extent, pipelines are perhaps uniquely vulnerable. No other energy transportation mode moves so much energy, over such a great distance, in a continuous stream whose continuity is so critical an aspect of its importance.[68]

Electrical power stations and transmission lines are particularly vulnerable to terrorist disruptions. Thirty years ago, the Congress's Joint Committee on Defense Production warned that large generating stations serving the nation's major metropolitan areas "present a relatively compact and especially inviting set of targets for a saboteur, a terrorist or an attacker."[69] Nor is the threat simply of a hypothetical nature. Attacks on power plants have become a standard part of terrorist and guerrilla warfare activity in countries like Afghanistan, El Salvador, and Cyprus, and even in G-7 nations like Italy and Britain.[70]

Since September 11th, government officials responsible for homeland security have been nervous about security in and around nuclear power plants. The cooling towers, containment buildings, and other facilities were designed many years ago to withstand a direct hit by smaller aircraft. No one knows for sure what the impact might be if terrorists were to strike a nuclear facility with a 747 jumbo jet. There is also concern that terrorists might target the spent radioactive fuel that lies relatively unprotected in storage pools at nuclear power sites. A major release of radioactivity from a nuclear power plant could cause 60,000 delayed cancer deaths, 60,000 genetic birth defects, 450,000 thyroid nodules, contamination of 5,300 square miles of land, and hundreds of billions of dollars of property damage over an extended period of time.[71]

Transmission lines are an equally inviting terrorist target. Forty years ago, the Defense Electric Power Administration warned that

> main transmission lines are extremely difficult to protect against sabotage as they are widespread over each state and traverse remote rugged and unsettled areas for thousands of miles. While these facilities are periodically patrolled, ample time is available for a saboteur to work unobserved.[72]

U.S. government officials, especially at the Pentagon, the Department of Energy, and the Department of the Interior, are well aware of the vulnerabilities but not sure what can be done, short of creating an alternative energy regime and a radically different electrical grid to go along with it. In lieu of a new energy vision for the country, we remain increasingly vulnerable to terrorist attacks in the years ahead. One Department of the Interior official summed up the enormity of the problem, acknowledging that "a relatively small group of dedicated, knowledgeable individuals . . . could bring down [the power grid supplying] almost any section of the country."[73]

OUR FOSSIL-FUEL CIVILIZATION is now vulnerable on so many different fronts that some close observers worry that a collapse of the system is no longer inconceivable. The global peak in oil production could occur sometime within the next two decades—and some geologists say much sooner—with natural gas peaking shortly thereafter.

The Middle East, which has most of the remaining oil and gas reserves, is becoming more politically unstable with each passing day as young Islamic militants challenge autocratic rulers and dictatorial governments. Muslim terrorists continue to pose a serious threat to homeland security in the U.S. and other G-7 countries, increasing the military costs of securing a dwindling reserve of oil and natural gas from the Persian Gulf. Meanwhile, anticipating the prospects of a diminishing supply of cheap crude oil and a dramatic rise in the price of oil on world markets, energy companies are beginning to turn back to coal, as well as ahead to previously untapped reserves of tar sand and heavy oil, in the hope of creating sufficient synthetic fuel to make up for a growing deficit in crude oil and natural gas. The switch to dirtier fuels, however, is likely going to mean even greater emissions of carbon dioxide into the atmosphere, further exacerbating the problem of rising temperatures on Earth during the remainder of the 21st century.

We are living through the senescent stages of a mature energy regime, with all the problems that go with it. The energy stocks are diminishing, and the entropy bill is mounting. The vast energy infrastructure we created to harness oil is growing old and becoming increasingly prone to disruption and disrepair. We are paying more and more to maintain the infrastructure while receiving less net energy benefits in return. At the same time, all of the economic and social institutions that come with a high-energy fossil-fuel regime are now threatened. Our petrochemical-based agriculture that supports a rising non-farm population and an urban way of life for more than half the people on Earth is particularly vulnerable. Agriculture is likely to be among the first casualties of a peak in global oil production. Dwindling reserves and ever-higher prices for oil could force a contraction in farm production all over the world, and with it a collapse of the manufacturing, service, information, and experiential economies built on top of it.

Depleted energy reserves could result in rolling electricity blackouts within a few years, especially in the world's major urban areas, with potentially devastating disruptions to our way of life. Loss of the electrical grid would inevitably force a partial migration out of uninhabitable metropolitan areas and back to rural regions.

Many skeptics will no doubt say that we have heard all of this before. Doomsayers cried wolf in the 1970s and early 1980s, predicting that the world would soon run out of oil, and their prophecies proved wrong. Why should these new predictions be any more believable? The doubters have a point. We may not be running short of cheap crude oil, at least not in the foreseeable future. And the predictions of rising temperatures on Earth could be equally misbegotten. The rising tide of Muslim militancy and the prospect of a worldwide intensification of terrorism aimed at the U.S. and other wealthy nations, including the oil-rich countries in the Middle East, could be exaggerated—and, even if they are real, they might quickly fizzle. The complex high-energy infrastructure and the increasingly centralized economic infrastructure built alongside it might not be as vulnerable to external threats and internal disruptions as some suggest.

Credible arguments can be mustered on both sides of this debate. Still, the accumulating anecdotal and statistical evidence makes a very strong case for those who argue that the oil-energy regime is moving into a sunset phase. It's hard to ignore the signs: Marginal returns have set in all along the line, the entropy bill is inexorably rising, the infrastructure itself is becoming increasingly stressed from the demands put on it, and the overall system is more and more vulnerable to disruptions of all kinds.

Perhaps the most compelling evidence indicating that we may be closing in on the final decades of the oil era has come from statistical analysis of the rise and fall of energy per capita in the world. A number of scientists, including Robert Romer, John Gibbons, and Richard Duncan, as well as BP, have plotted the worldwide bell curve of energy production per capita. Their findings are remarkably similar and provide irrefutable evidence that the oil age is winding down as fast as it revved up.

According to BP, world oil production per capita peaked in 1979 and has been declining ever since. That is because while oil production has been increasing, the world population has been growing even faster.[74]

Even the optimists cannot explain away the long-term decline in oil production per capita. There is no foreseeable scenario on the horizon in which oil—or, for that matter, natural gas, coal, heavy oil, tar sand, or nuclear power—can reverse the trend and provide enough energy per capita to bring the growing world population back up to the peak reached in 1979. As far as energy per capita is concerned, the human race is moving inexorably down the backside of the energy bell curve for the oil era. That being the stark reality, the single most important question now facing human civilization is whether or not a new energy regime can be found and harnessed in time to replace fossil fuels and meet the needs of a growing human population on Earth in the coming century.

8

THE DAWN OF THE HYDROGEN ECONOMY

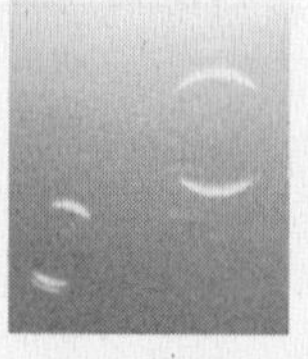

In 1874, Jules Verne, the popular science fiction writer, published a curious book entitled *The Mysterious Island*. The book was about the adventures of five Northerners during the American Civil War who were thrown off-course in a balloon while escaping a Confederate encampment. They eventually landed 7,000 miles away on a little island. One day they were musing about the future of the Union, and one of the members of the party, a sailor named Pencroft, asked the engineer, Cyrus Harding, what might happen to commerce and industry if America were to run out of coal. "And what will they burn instead of coal?" asked Pencroft. "Water," exclaimed Harding, to everyone's surprise. Harding went on to explain:

> Water decomposed into its primitive elements and decomposed doubtless, by electricity, which will then have become a powerful and manageable force.... Yes, my friends, I believe that water will one day be employed as fuel, that hydrogen and oxygen which constitute it, used singly or together, will furnish an inexhaustible source of heat and light, of an intensity of which coal is not capable.... Water will be the coal of the future.[1]

One hundred twenty-seven years after Verne first hinted of a future where all the energy needs of civilization would come from hydrogen extracted from water, Phil Watts, the chairman of Royal Dutch Shell, delivered an address about the future of energy at a forum sponsored by the United Nations Development Program. The setting was New York City, just three weeks after the terrorist attack on the World Trade Center. With the smell of toxic smoke from ground zero still heavy in the air over Manhattan, Watts's thoughts turned to the future of energy. He informed his audience that Shell was preparing for the "End of the Hydrocarbon Age." In the 21st century, said Watts, coal, oil, and natural gas, the great fossil fuels that had propelled the world into the industrial era, would give way to a revolutionary new energy regime based on hydrogen and that Shell had already committed up to a billion dollars to making the transition into a renewable resource economy.[2]

Verne's vision of a hydrogen future is now the subject of feverish attention in the corporate boardrooms of the world's leading energy companies, automotive companies, and utilities, as well as among policymakers and a growing number of nongovernmental organizations, both in the advanced industrial nations and the Third World.

Decarbonization

Hydrogen is the most abundant element in the universe. It makes up 75 percent of the mass of the universe and 90 percent of its molecules.[3] Effectively harnessing it as a source of power would provide humanity with a virtually unlimited source of energy—the kind of energy elixir that has long eluded alchemists and chemists alike. In a way, Jules Verne's premonition of a hydrogen future was already becoming apparent by the last quarter of the 19th century. In less than a century, fuel wood had given way to coal, and coal was being challenged by a new upstart, oil. The "decarbonization" of energy that would inevitably lead to a hydrogen future was already well under way.

"Decarbonization" is a term scientists use to refer to the changing ratio of carbon to hydrogen atoms with each succeeding energy source. Fuel wood, which for most of history was humanity's primary energy fuel, has the highest ratio of carbon to hydrogen atoms, with ten carbon atoms per hydrogen atom. Among fossil fuels, coal has the highest carbon-to-hydrogen ratio, about one or two carbon atoms to one hydrogen atom. Oil has one carbon atom for every two hydrogen atoms, while natural gas has only one carbon atom to four hydrogen atoms. This means that each successive energy source emits less CO_2 than its predecessor. Nebojsa Nakicenovic, of the International Institute for Applied Systems Analysis in Vienna, estimates that the carbon emission per unit of primary energy consumed globally has continued to fall about 0.3 percent per year over the past 140 years.[4]

Of course, because of the sheer volume of coal and oil being burned, CO_2 emissions have nonetheless continued to rise during that time, increasing the Earth's surface temperatures. While the current shift from coal and oil to natural gas promises to reduce the amount of CO_2 emissions per unit of energy burned even further, the volume of natural gas used will continue to mean more CO_2 emissions and rising temperatures on Earth, but not as much as would be the case if we were still relying primarily on coal or oil. Jesse Ausubel, senior research associate at the Rockefeller University in New York City, sums up the historic significance of the energy path the world has journeyed down:

> [T]he most important, surprising, and happy fact to emerge from energy studies is that for the last 200 years, the world has progressively favored hydrogen atoms over carbon. . . . The trend

toward "decarbonization" is at the heart of understanding the evolution of the energy system.[5]

Hydrogen completes the journey of decarbonization. It contains no carbon atoms. Its emergence as the primary energy source of the future signals the end of the long reign of hydrocarbon energy in human history. Hydrogen, the power source of the sun, (it makes up 30 percent of the mass of the sun), is being increasingly looked to as the great hope for humanity's continued advance on Earth.[6] It is the lightest and most immaterial of all forms of energy and the most efficient when burned.

The steady progression from heavy to light and from material to immaterial in our forms of energy has paralleled, at every step of the way, the increasing weightlessness of industrial activity, from the onset of heavy steam-age technologies of early industrial capitalism to the light and virtual information-age technologies of the 21st century. Indeed, dematerialization of energy and dematerialization of economic activity invariably go hand in hand. Decarbonization has meant not only the steady elimination of carbon atoms but, with it, the dematerialization of energy—from solids (like coal) to liquids (oil) and now to gases (both natural gas and hydrogen). The shift in energy from solids to liquids to gases makes for a faster and more efficient energy throughput—oil travels faster through pipelines than coal is transported over rail, and gas travels even faster and more lightly through pipelines than liquid oil—and gives rise to like-minded technologies, goods, and services that are increasingly fast, efficient, light, and virtual.

Speaking before the Science Committee of the U.S House of Representatives in April 2001, Texaco executive Frank Ingriselli drew the connection between the great shifts taking place in the global economy and society and the dawn of the hydrogen era. He observed that "greenery, innovation, and market forces are shaping the future of our industry and propelling us inexorably toward hydrogen energy" and warned that "those who don't pursue it, will rue it."[7]

Hydrogen is ubiquitous here on Earth and is found in water, fossil fuels, and all living things. The hydrogen in water and organic forms comprises 70 percent of the Earth's surface.[8] However, as we will shortly see, it rarely exists free-floating and alone, as do coal, oil, and natural gas. It is an energy carrier, a secondary form of energy that has to be produced, like electricity.

The Energy Elixir

Hydrogen was first discovered by the British scientist Henry Cavendish. In a paper delivered before the Royal Society of London in 1776, he reported on an experiment in which he had produced water by combining oxygen and hydrogen with the aid of an electrical spark. Since the elements had not yet been named, he called one "life-sustaining air" and the other "inflammable air."[9] The French chemist Antoine Laurent Lavoisier successfully repeated Cavendish's experiment in 1785 and called the "life-sustaining air" oxygen and the "inflammable air" hydrogen.[10]

The first practical use of hydrogen, not surprisingly, was in warfare. A chemist, Guyton de Norveau—who was also a member of the Committee for Public Salvation, one of the warring factions in the hostilities that had broken out in France after the storming of the Bastille in 1793—made the suggestion that hydrogen gas be produced in large quantities and be used in reconnaissance balloons. The first hydrogen generator was built at an army camp just outside Paris in 1794.[11]

Hydrogen was being commercially produced in the 1920s in Europe and North America. Canada's Electrolyser Corporation Limited led the way. The company, which in the early days was called the Stuart Oxygen Company, manufactured and sold the first commercial electrolizers to a U.S. company in San Francisco in 1920. Electrolizers are machines that split water into hydrogen and oxygen. Today, Electrolyser Corporation is one of the largest manufacturers of electrolytic hydrogen plants in the world.[12]

The first scientist of note to envision the full potential of hydrogen was John Burden Sanderson Haldane, who later went on to become one of the distinguished geneticists of the 20th century. In

1923, and still in his twenties, Haldane gave a lecture at Cambridge University in which he predicted that hydrogen energy would be the fuel of the future. He then went on to outline, in a scientific paper, the case for hydrogen and how it would be produced, stored, and used. His notion was so revolutionary at the time that it was met with incredulity by his peers within the academy. Yet, in its every particular, his thesis was tantamount to a working blueprint of how hydrogen would later be harnessed and exploited.

Haldane began with a spirited defense of the superiority of hydrogen over other forms of energy. He wrote that "liquid hydrogen is weight for weight the most efficient known method of storing energy, as it gives about three times as much heat per pound as petrol."[13] Haldane then turned to the issue of how hydrogen would be produced. He predicted that, in four centuries, Britain's energy requirements would be met by

> rows of metallic windmills working electric motors which in their turn supply current at a very high voltage to giant electric mains. At suitable distances there will be great power stations where during windy weather the surplus power will be used for the electrolytic decomposition of water into oxygen and hydrogen. These gases will be liquefied and stored in vast vacuum jacketed reservoirs probably sunk in the ground. . . . In times of calm the gases will be recombined in explosion motors working dynamos which produce electrical energy once more, or probably in oxidation cells. These huge reservoirs of liquefied gases will enable wind energy to be stored so that it can be expended for industry, transportation, heating and lighting as desired.[14]

Haldane even anticipated the obstacles in the way of making the transition to a hydrogen energy regime, as well as the far-reaching social and environmental consequences that would result. On the first score, he acknowledged that "the initial costs will be very considerable but the running expenses less than those of our present system." The great social advantage of adopting a hydrogen energy regime is that "energy will be as cheap in one part of the country as another, so that industry will be greatly decentralized."[15] The environmental benefits, noted Haldane, would be equally attractive as "no smoke or ash will be produced."[16]

Hydrogen was first exploited as a fuel in aviation in the 1920s and '30s. German engineers used it as a booster fuel in the Zeppelins (the German dirigibles that ferried commercial passengers across the Atlantic). The ships' primary fuel was a mixture of benzene and gasoline. The engines were modified so that the normal blow-off of hydrogen that was used to maintain the buoyancy of the airship could be used as extra fuel.[17] By the 1930s and '40s, hydrogen was being used in Germany and England as an experimental fuel for automobiles, trucks, locomotives, and even submarines and trackless torpedoes.

Today, some 400 billion cubic meters of hydrogen are produced globally—equivalent to approximately 10 percent of world oil production in 1999.[18] Much of the hydrogen is used as a chemical raw material for the production of such things as ammonia-based fertilizers and in the hydrogenation of edible organic oils made from soybeans, fish, peanuts, and corn. Hydrogen is also used to convert liquid oil into margarine. Hydrogen is used in the process of manufacturing of polypropylene and is used to cool generators and motors.[19]

While hydrogen has been widely used in the refining process as a feedstock and raw material for the production of a range of products, its value as a fuel was largely ignored in the post–World War II era, despite early experimental successes in the 1920s and '30s in aviation and automobile transport. It was not until the oil crisis in 1973 that scientists, engineers, and policymakers decided to take a second look at hydrogen as an all-purpose form of energy. In that year, the first International Conference on Hydrogen was held in Miami Beach, and the International Association for Hydrogen Energy was established, along with a monthly journal, the *International Journal of Hydrogen Energy*. A small group of enthusiasts, who called themselves the "Hydrogen Romantics," began to proselytize within the energy industry, hoping to win over converts to the hydrogen vision. T. Nejat Veziroglu, president of the association and a member of the group, summed up the excitement at the time, exclaiming that hydrogen "was a permanent solution to the depletion of conventional fuels, [and a] permanent solution to the global environmental problem."[20]

In the years that followed, the U.S. and other governments began to invest small amounts of public funds in hydrogen research. The U.S. program never exceeded $24 million.[21] The European Economic Community spent between $72 million and $84 million on hydrogen research in the 1970s. As the energy crisis waned and the price of oil began to drop again on world markets in the 1980s, government funding for hydrogen research declined significantly.

Interest in hydrogen began to pick up again in the 1990s after the publication of alarming studies and reports warning that increased CO_2 emissions from the burning of fossil fuels was heating up the planet and posing a potentially grave threat to the Earth's biosphere. A growing number of scientists began to raise the possibility of making a transition from hydrocarbon fuels to hydrogen as a way to address the problem of global warming. Decarbonization became the rallying point for geologists, climatologists, and environmentalists alike. Meanwhile, experimental research, both in academia and the commercial sector, began to lay the technical foundation for a hydrogen future.

The Soviet Union successfully converted a passenger jet to run partially on liquid hydrogen in 1988, and that same year an American, William Conrad, became the first person to fly an airplane fueled only by liquid hydrogen. In 1992, the Fraunhofer Institute for Solar Energy Systems in Germany created the first solar home, using hydrogen for long-term energy storage. The next year, Japan committed $2 billion to a thirty-year plan to promote hydrogen energy around the world. In 1994, the first hydrogen-powered buses took to the streets in Geel, Belgium. The Chicago Transit Authority began testing its own hydrogen-fueled buses a year later. The Royal Dutch/Shell Group made its first tentative steps into the hydrogen era in 1998, setting up a "Hydrogen Team" to explore business ventures, and a year later it set up a hydrogen division.

These milestones and countless other piecemeal efforts over the past century have helped raise hydrogen's profile. It wasn't until 1999, however, that the full potential impact of hydrogen energy became apparent. In February of that year, Iceland announced an ambitious and daring long-term plan to become the first hydrogen economy in the world.[22]

A joint venture made up of three transnational companies (the

Royal Dutch/Shell Group, Daimler-Chrysler, and Norsk Hydro) and six Icelandic participants (the Reykjanes Geothermal Power Plant, the Reykjavik Municipal Power Company, a fertilizer plant, the University of Iceland, the Iceland Research Institute, and the New Business Venture Fund) will be responsible for implementing the plan. The Icelandic participants control 51.01 percent of the venture.[23]

Thorsteinn Sigfusson, a professor of physics at the University of Iceland in Reykjavik and chairman of Iceland New Energy, says that the consortium's goal is to run the entire Icelandic economy on hydrogen within twenty years, virtually eliminating fossil-fuel energy from the country. The plan first calls for converting the country's fleet of cars, buses, trucks, and trawlers to hydrogen and, shortly thereafter, for using hydrogen to generate electricity and provide heat, light, and power for its factories, offices, and homes. Iceland is already being referred to as "the Bahrain of the North," and there is talk about eventually exporting hydrogen from there to Europe, making the country the first hydrogen-producing nation.[24]

A similar project is now under way in the state of Hawaii. The state, which imports much of its oil, mainly from Asia and Alaska via tankers, hopes to become energy self-sufficient by harnessing its abundant geothermal and solar energy and converting it to hydrogen fuel. The state legislature approved a small grant in April 2001 to help create a public/private partnership to exploit hydrogen energy. The University of Hawaii received an additional $2 million grant from the U.S. Department of Defense to advance the hydrogen agenda. Hermina Morita, a state representative who chairs the legislative committee to reduce Hawaii's dependence on oil, says that Hawaii's long-term goal is "to be capable of producing more hydrogen than we need, so we can send the excess to California."[25]

It is worth noting that General Motors, the largest automaker in the world, was the first to use the phrase "the hydrogen economy."[26] That was back in 1970, when GM engineers began to look at hydrogen as the possible energy fuel of the future. Thirty years later, after many trailblazing efforts had begun to establish the viability of a hydrogen future, GM's Executive Director of Advanced Technology Vehicles, Robert Purcell, speaking at the annual meeting of the National Petrochemical and Refiners Association in May

2000, told the members that "our long-term vision is of a hydrogen economy."[27]

Producing Hydrogen Energy

As mentioned earlier, hydrogen is found virtually everywhere but rarely stands alone in nature. Instead, it is embedded in water, fossil fuels, and all living things, and must be extracted to be used as a form of energy.

There are a number of ways to produce hydrogen. Today, nearly half the hydrogen produced in the world is derived from natural gas via a steam-reforming process. The natural gas reacts with steam in a catalytic converter. The process strips away the hydrogen atoms, leaving carbon dioxide as the by-product. Coal can also be reformed through gasification to produce hydrogen, but this is more expensive than using natural gas. Hydrogen can also be processed from oil or gasified biomass.

While using steam to reform natural gas has proven the cheapest way to produce commercial hydrogen, natural gas is still a hydrocarbon and emits CO_2 in the conversion process. Proponents argue that, in the future, the CO_2 that is generated in the conversion process could be isolated and sequestered in underground storage sites, including depleted oil or gas fields and deep coal beds, although they acknowledge that this would increase the costs of producing the hydrogen. The feasibility of sequestration technology is still in doubt, and even proponents say that at best the commercial application is at least ten years away.[28]

Most industry analysts are convinced that, in the foreseeable future, natural gas and, to a lesser extent, other fossil fuels will be the primary sources of hydrogen. However, their analysis hinges on the risky assumption that enough cheap natural gas exists to meet not only the increasing demands for hydrogen but also the increasing demands of the electrical power industry, in which a new generation of gas-fired plants is coming on line and is expected to provide much of our new electricity, at least in the U.S., in the coming decades. If, however, global peak production of natural gas occurs by 2020, as some geologists are now forecasting, then other ways of producing hydrogen will need to be found. Even the Electric Power Research Institute (EPRI), the think tank for the American Utilities Industry, says in its own internal study that it may not be possible to generate enough natural gas at cheap prices to sustain the projected increases in electrical generation being projected, even before factoring in a significant increase in the use of natural gas to produce hydrogen. The EPRI study forecasts that natural-gas generation will increase from 15 to 60 percent over the next twenty years with the introduction of hundreds of new gas-fired generation plants. Despite the fact that power companies are already locked into gas-fired generating plants, EPRI's Dr. Gordon Hester says that, based on their study, "reliance on natural gas for electricity generation could not be maintained at such a high level for an extended period of time."[29] Increasing demand for electricity, according to the institute, will likely cause the price of natural gas to rise, encouraging the shift to less-costly fuels that produce fewer emissions. The result, according to the EPRI study, would be that the use of natural gas to generate electricity would decline significantly after 2025.[30]

If natural gas will not be available in sufficient volume to meet the demands for electricity twenty years from now, then relying on it as a source of producing freestanding hydrogen seems misplaced. There is, however, another way to produce hydrogen without using fossil fuels in the process. Electrolysis, as mentioned earlier, uses electricity to split water into hydrogen and oxygen atoms. The process has been around for more than 100 years. Here is how it works: Two electrodes, one positive and the other negative, are submerged in pure water that has been made more conductive by the addition of an electrolyte. When electricity—direct current—is applied, the hydrogen bubbles up at the negatively charged electrode (the cathode) and oxygen at the positively charged electrode (the anode).

Industrial electrolysis plants exist in a number of countries. The equipment includes the basic tank as well as an electric power converter to change alternating current to direct current, pipes to carry hydrogen and oxygen away from the cells, and equipment to dry the gases after they are separated from the electrolyte.[31]

Electrolysis has not been widely used—only 4 percent of the hydrogen produced annually is derived from electrolysis of water—because the cost of the electricity used in the process makes it uncompetitive with the natural-gas steam-reforming process. The electricity can cost three to four times as much as steam-reformed natural-gas feedstock.[32] This last point needs to be emphasized, because many observers have come to believe that the electrolyzer process itself is expensive and inefficient, when in fact it is the cost of generating electricity in large, centralized power plants that makes the process so costly. The Institute of Gas Technology reports that "most commercial electrolyzers available today are capable of electricity-to-hydrogen efficiencies above 75 percent, while their capital-cost potential is far less than that of power stations that would be required to run them."[33]

The real question, then, is whether it is possible to use renewable forms of energy that are carbon-free, like photovoltaic, wind, hydro, and geothermal, to generate the electricity that is used in the electrolysis process to split water into hydrogen and oxygen. A growing number of energy experts say yes, but with the qualification that the costs of employing renewable forms of energy will need to come down considerably to make the process competitive with the natural-gas steam-reforming process. Seth Dunn, of the World Watch Institute, makes note of the fact that, while "the costs of PV- and wind-based electrolysis is still high," they are "projected to be cut in half over the next decade."[34] If natural-gas production peaks, sending prices dramatically higher, a point could be reached where using renewable sources of energy to produce electricity for the electrolysis process would be cheaper.

To harness the sun directly and turn it into useful energy has long been the dream of scientists and engineers. The amount of energy potentially available from the sun's rays is truly incredible. John Houghton tells us that "as much energy arrives at the Earth from the sun in forty minutes as we use in a whole year."[35] Capturing that sun was only a far-off vision 100 years ago. No longer. Photovoltaic devices (PVs), using semiconductor material to convert sunlight into electricity, are being installed all over the world, and, while they are still expensive, their costs are coming down. The cost of solar cells has dropped 95 percent since the 1970s.[36]

PVs are already being used commercially as power sources for watches and calculators. Spacecraft use "solar panels" or possess "solar arrays" covered by PV cells, providing electric power for astronauts working in outer space. PV efficiency is between 10 and 20 percent, and a one-square-meter panel of cells produces between 100 and 200 watts of electrical power.[37]

The first successful large-scale harnessing of solar energy for electricity generation came in the 1980s. Nine solar thermal power plants using parabolic trough mirrors to capture the sun's rays were built in the Mojave desert between Las Vegas and Los Angeles. They provide 354 megawatts of electricity to homes and industry in the region.[38]

Photovoltaics have also grown in popularity in the developing countries in the Southern Hemisphere in recent years. BP Solar, a company that produces 10 percent of the photovoltaic cells in the world, has put together a $48 million solar project on the island of Mindanao in the Philippines. The project, the largest solar-energy initiative in the world, when completed, will provide electricity for 400,000 residents of 150 villages in one of the poorest areas in the Pacific. The electricity will also power sixty-nine irrigation systems and ninety-seven drinking-water systems and will provide power and light for dozens of schools and medical clinics.[39]

PV power is still two to five times more expensive than conventional electricity generated from fossil fuels, but the costs continue to go down with new technological innovations and greater economies of scale. In 1998, PV prices dropped to under $4 per watt for the first time.[40]

On September 26, 1995, the first solar hydrogen facility began operations in the United States in El Segundo, California. The $2.5 million project, a joint effort by Clean Air Now (CAN), a local environmental organization, and the Xerox Corporation, captures solar radiation using an advanced photovoltaic system designed by Solar Engineering Applications Corporation, a California high-tech firm, and converts it into electricity, which is then fed into an electrolyzer manufactured by the Electrolyser Corporation of Canada. The electrolyzer produces between 1,500 and 2,000 standard cubic feet (scf) of hydrogen per day. After excess water vapor has been removed, the hydrogen gas is compressed to 5,000 pounds per square inch (psi), dried, and stored. Hydrogen gas from the facility is used to fuel the Ford Ranger trucks that have been modified to run on the decarbonized fuel.[41]

Today, companies like Royal Dutch/Shell and British Petroleum are investing billions of dollars in solar and other renewable energy technologies in anticipation of the phaseout of fossil fuels. Shell believes that solar and other renewable energies will make up more than one-third of the market for new electricity generation by 2050, with global sales topping $150 billion.[42] Shell recently joined with Siemens to create the fourth-largest solar energy company in the United States.[43] John Browne, the CEO of British Petroleum, is even more bullish, forecasting that 50 percent of the total energy demand in the world will be met by solar and other renewable resources by 2050.[44]

Two percent of solar energy is converted into wind energy through atmospheric circulation.[45] Wind is now the most cost-effective form of renewable energy. A wind energy generator is comprised of two-or three-bladed propellers approximately fifty meters in diameter. In an area where average wind speed is 7.5 meters per second, the propellers will generate about 250 kilowatts of electricity. According to the American Wind Energy Association, the cost of a kilowatt-hour of wind-generated energy has dropped from forty cents in the early 1980s to less than five cents and, in some locales, as low as three cents. The U.S. Department of Energy says that, in some areas of the country, wind power is now so cheap and efficient that it is competitive with gas-fired power plants.[46] When wind power drops to 1.5 cents per kilowatt-hour in the next few years, electrically generated hydrogen, using wind power, will be competitive with gasoline.[47]

In recent years, worldwide wind capacity has been growing at an annual rate of 27.75 percent.[48] The European Wind Association predicts that wind energy can produce 10 percent of worldwide electricity by 2020.[49] The wind industry is one of the fastest-growing market segments in the world economy. Analyst Michael Kujawa of Allied Business Intelligence Inc. estimates that wind-turbine sales will exceed $200 billion by 2010.[50] Europe is leading the way into wind energy. Of the fifteen gigawatts (GW) of wind capacity that currently exist in the world, ten GW have been installed in Europe.[51] In Denmark, wind is now providing 14 percent of the country's total electricity output.[52] In some of the northern regions of Germany, wind power makes up 15 percent of the generated electricity.[53] The high winds on the west coast of Britain make wind energy a strong contender in the country's future energy mix. Some forecasters predict that wind energy could supply 10 percent or more of Britain's electricity needs in the near future.[54] A study prepared by Germanischer Lloyd and Gerrad Hassan estimates that the wind-generating potential along the coastal regions of the Baltic and North Seas could produce enough wind to provide the electricity needs of the entire European continent.[55]

Developing countries are also expanding their wind-generating capacity. India is one of the world's top-five wind producers, having already installed one GW. By 2030, India could be producing ten GW of electric power with wind technology—equivalent to 25 percent of its current needs.[56]

In the U.S., the plains states stretching from Texas to the Dakotas are being looked to as prime wind-generating regions. In Texas, a wind farm is currently being constructed that will provide electricity for 139,000 homes.[57]

Hydroelectric power is still another potential source of renewable energy for providing hydrogen. Nearly 20 percent of solar energy is used to evaporate water on the surface of the Earth. When the water vapor condenses and falls as precipitation, it makes hydropower possible.[58] Hydropower currently makes up 10 percent of the electrical power generated in the United States and 19 percent of the world's electricity generation.[59] Seth Dunn suggests that other countries with cheap hydroelectric power, like Brazil, Canada, Norway, Sweden, and Iceland, may be among the first to use renewable electrolysis on a large scale.[60]

Geothermal energy, although not yet widely exploited—it makes up only 0.1 percent of total world energy—also has potential as a renewable energy resource.[61] Hot water and steam, which lie deep beneath the Earth's surface in volcanic rock, geysers, and hot springs, can be converted to electricity. According to the U.S. Department of Energy, the geothermal resources in the U.S. alone are estimated to exceed 70,000,000 quads.[62] That is enough potential energy to supply power for human consumption for hundreds of thousands of years. Iceland and Hawaii, in the quest to become hydrogen economies, are going to rely increasingly on their geothermal power to generate the electricity needed to produce hydrogen from water. Geothermal "provinces" exist in the Pacific Ocean, India, southeastern Asia, along the shores of China and Japan, on the west coasts of Canada, the U.S., and South America, and in parts of the Mediterranean, Russia, and eastern Africa.[63]

Biomass, in the form of agricultural and industrial waste, can also be used to generate electricity to electrolyze water and produce hydrogen. The United Kingdom, for example, produces thirty million tons of solid waste a year. If this waste were incinerated for power generation, it could produce enough electricity to provide 5 percent of the U.K.'s total requirement.[64] According to the U.S. Department of Energy, the gasification of biomass, using a new generation of advanced turbines, could reduce the cost to 4.5 cents per kilowatt-hour over the next few years.[65] Shell forecasts that, by the end of the first decade of the 21st century, energy produced by gasification of biomass could provide 5 percent of the world's energy needs.[66] While biomass produces CO_2 in the process of gasification, the CO_2 is recaptured with new plant growth, making the biomass carbon-neutral.

The most important aspect of using renewable resources to produce hydrogen is that the sun's energy, and wind, hydro, and geothermal energies, will be convertible into "stored" energy that can be applied in concentrated forms whenever and wherever needed, and with zero CO_2 emissions. This point needs to be emphasized. A renewable energy future is made far more difficult, if not impossible, without using hydrogen as a means for energy storage. That's because when any form of energy is harnessed to produce electricity, the electricity flows immediately. So if the sun isn't shining, or the wind isn't blowing, or the water isn't flowing, or fossil fuels are not available to burn, electricity can't be generated and economic activity grinds to a halt. Hydrogen is one very attractive way to store energy to ensure an ongoing and continuous supply of power for society.

Creating an infrastructure to store hydrogen, however, raises additional cost questions. Proponents of the renewable energy/hydrogen complex are pinning their hopes on breakthroughs being made in the development of small stationary and portable fuel cells and on the fast-growing market for them as mini-plants for use in factories, offices, retail stores, homes, and automobiles.

Fuel Cells—The Mini-Power Plants

Fuel cells are not a new invention. In fact, they predate the internal combustion engine. But they were not of much commercial interest until the 1960s, when NASA made the decision to use them in the space program to provide electrical power for spacecraft. The Apollo lunar mission used fuel cells developed by the Pratt & Whitney division of the United Aircraft Corporation. The company, which later became United Technologies, supplied fuel cells for the space-shuttle program as well.[67]

Fuel cells are like batteries—but with one big difference. Batteries store chemical energy and convert it to electricity. When the chemical energy runs out, the battery is discarded. Fuel cells, by contrast, do not store chemical energy. Instead, they convert the chemical energy of a fuel that is fed into them to generate electricity. They do not require recharging and will continue to generate electricity as long as external fuel and an oxidant are supplied.

Fuel cells require hydrogen fuel. The hydrocarbon fuels are too "dirty" to be used as primary fuels for the cell. A fuel cell is made up of a negatively charged anode on one side, a positively charged cathode on the other, and an electrolyte in the middle that is made up of an alkaline or watery acidic solution or a plastic membrane, allowing the electrically charged hydrogen atoms to travel from the anode to the cathode. Commercial fuel cells are composed of many individual cells stacked atop one another. Hydrogen is fed into the anode side of the cell, where a chemical reaction splits the hydrogen atom into a proton and an electron. Freed electrons exit through the external electrical circuit in the form of direct-current electricity. The hydrogen ions (the proton) travel through the electrolyte layer to the positively charged cathode. The flow of electrons returns to the cathode, where they react with hydrogen ions and oxygen in the air to form water.[68] Fuel cells work using a process that is the reverse of electrolysis. They have no moving parts, are silent, and are up to two-and-one-half times more efficient than internal combustion engines.[69] And the only effluents are electricity, heat, and pure distilled water.

Fuel cells powered by hydrogen could potentially produce enough electricity to serve the needs of the human race far into the future. Moving beyond the fossil-fuel era, however, will not be easy. It is still expensive to produce hydrogen. Moreover, at present, most fuel cells use natural gas and other fossil fuels as fuel stock. As mentioned, some hydrogen is already being produced on a small scale by the harnessing of solar, wind, hydro, geothermal, and other renewable forms of energy to produce the electricity needed to separate out hydrogen from water. In recent years, solar- and wind-power–based electrolysis systems have been set up in Germany, Italy, Spain, Switzerland, Finland, the United States, and even Saudi Arabia.[70] Still, the process is more expensive than the steam reforming of hydrocarbon fuels.

Fuel cells are also expensive. Like any new technology, the production of fuel cells has not yet reached the critical threshold where economies of scale kick in to significantly reduce the cost per manufactured unit. Still, dozens of startup companies, as well as some of the biggest transnational companies in the world, are entering the field in hopes of leading the way to a hydrogen economy. Ballard Power Systems of Burnaby, British Columbia, and Plug Power of Latham, New York, have both launched ambitious marketing plans to outfit commercial businesses and homes with stationary fuel-cell power plants. The home units are 1- to 15-kilowatt power plants, while the commercial units are 60 to 250 kilowatts. Plug Power is partnering with General Electric and expects to be on the market in the latter part of 2002 with thousands of residential units.[71]

Distributed Generation

Virtually all of the players in the new energy game are looking to a radical new way to deliver electricity, called distributed generation (DG), to address the cost questions and pave the way to a new energy era. DG turns the conventional logic of how to deliver electricity on its head. For most of the 20th century, electrical power was generated at huge power plants and then transported over long distances to end users via transmission lines. Centralized power created an economy of scale, which made the production and distribution of electricity relatively cheap. The high capital cost in building mega-power plants and vast transmission grids could only be recouped by allowing companies to control the whole of a regional market. So, in the U.S. and elsewhere around the world, electrical power was either constituted as a publicly owned utility or as a private utility regulated by the government as a natural monopoly.

By the 1970s and '80s, however, the centralized power-generating infrastructure was coming under increasing attack by critics who argued that its very size was making it dysfunctional in addressing a host of new challenges, including the dramatic rise in the cost of energy resulting from the Arab oil embargo, the OPEC price spikes, and the growing problem of CO_2 emissions and other pollutants that were threatening the environment and public health. In response to increasing public pressure to find new ways to conserve energy, the U.S Congress enacted the Public Utilities Regulatory Policy Act (PURPA) in 1978, legislation designed, in part, to encourage the use of co-generation (recycling the waste heat from electricity generation for heating and power in factories and offices). The act encouraged new businesses to enter the energy field, providing the beginnings of market competition.

Meanwhile, natural-gas deregulation lowered prices for that fuel and stimulated new breakthroughs in gas-fired power generation. The new gas-fired turbines were cost effective at sizes of 100 megawatts or less, and they required far less up-front capital investment than the conventional 1,000-megawatt coal and nuclear plants. The new generation of cheap, gas-fired power plants also required less time to install and were easier to maintain. A new generation of independent energy producers argued that the long-standing rationale for granting power companies "natural monopoly" status—to wit, the high capital costs of providing commercial access to electricity—no longer held sway in light of the technical innovations that were making power generation both cheap and versatile.

Other problems plagued the utilities industry, contributing to growing pressure from both industry and the public to open the doors to new forms of competition and new ways of delivering electrical power. The power companies had invested heavily in nuclear power plants in the 1960s and '70s. By the 1980s, significant cost overruns and shutdowns were being passed on to consumers in the form of dramatic price increases for electricity in some locales. Strapped for funds, many of the nation's largest power companies failed to invest sufficiently in new generating capacity to meet increased commercial and residential demand. Power outages, brownouts, and blackouts became more frequent, angering commercial customers (who had to cover the losses for the downtime) as well as residential customers, who were not used to being inconvenienced. The upshot was that these natural monopolies found themselves with fewer friends willing to champion their special status at the very time that independent power producers were knocking at the door, anxious to win away their customers.

The deregulation fervor, which reached a peak during Ronald Reagan's and George Bush's presidential terms in the United States and Margaret Thatcher's and Helmut Kohl's terms as prime ministers in the United Kingdom and Germany, rocked every industry—but none was more deeply affected than energy. In 1992, the U.S. enacted the Energy Policy Act, opening up competition within the electric power industry. Independent power producers began to challenge the giants by introducing small-scale technologies serving niche markets. The era of distributed generation had dawned.

"Distributed generation" generally refers to integrated or standalone small electricity-generation power plants that are located near or at the site of the end user—factories, commercial businesses, public buildings, neighborhoods, and private residences.[72]

Today, the most popular micro-power technology is reciprocating engines that run on diesel fuel or natural gas. Gas turbines and micro turbines powered by fossil fuels are also becoming more widely used in the micro-power market. However, there is a growing consensus among experts in the field that, in the longer run, hydrogen-powered fuel cells will eventually assume dominance and become the energy leader in the distributed-generation market. Besides being more efficient than a combustion engine in generating electricity, as well as being less polluting, the fuel cell is also more flexible. Fuel cells come in modules, and this allows the end user to customize the unit to his or her current power-generating needs—and if more power is needed in the future, additional fuel-cell modules can be added at little additional expense.[73]

Electricity generated from fuel cells now costs $3,000 to $4,000 per kilowatt, whereas electricity generated from a typical gas-fired central power plant runs between $500 and $1,000 per kilowatt.[74] Of course, the price of generating electricity with fuel cells will continue to go down as more units are sold, and this will create greater economies of scale and spur more innovation. Industry analysts are bullish on the market potential for fuel cells and distributed generation, and they cite a number of reasons why they believe that the electrical grid is likely to shift from centralized power generation to decentralized power generation located at or near the end user.

First, there is the growing concern over brownouts and blackouts within industry and commerce, especially within the high-tech electronics, computer, and software fields. Businesses now talk about the need for "premium power." In manufacturing, banking, telecommunications, and other industries virtually dependent not only on an uninterrupted flow of electronic information over internets and intranets but also on maintenance of critical software databases and digital equipment of all kinds, the loss of electricity can mean significant losses in production and distribution as well as loss of critical knowledge-based assets.

A disruption in the flow of electricity for as little as eight-thousandths of a second can spell disaster. In 1997, a brief disruption in the supply of electricity at the National Bank of Omaha in Nebraska, caused a crash of the computer systems that dealt with the bank's major credit-card transactions. The bank estimates that a one-hour outage costs more than $6 million in lost revenue for the institution. The bank subsequently installed four 200-kilowatt fuel cells at its technology center to make sure that it would never again be vulnerable to a loss of electricity from the main power grid.[75]

In some manufacturing industries, even a momentary flickering in power can result in the shutdown of equipment and result in millions of dollars of losses. Hewlett Packard estimates that a fifteen-minute outage at one of its chip-fabricating factories cost the company $30 million—that is, equivalent to half the plant's total power budget for a whole year.[76] Power outages cost U.S. industry between $12 billion and $26 billion per year, and the figures are likely to rise as more and more companies become dependent on digital technologies, computer software, and electronic networks of all kinds in their commercial relationships.[77]

Even large commercial buildings are installing fuel cells so that they won't be vulnerable to the kind of power outages that have crippled cities like New York in the past. A new building recently completed in Times Square in New York City has 200-kilowatt fuel cells that provide all of its hot water, light its façade and supply backup power for the main grid.[78]

Premium power is a $7- to $10 billion-per-year market in North America, mostly in the form of backup power generation.[79] The market could expand dramatically. If companies continue to experience increasing power outages from the centralized power grids and rising prices for their electricity, they could opt to make their backup distributive-generating power plants their primary source of power.

Vital public services are also increasingly vulnerable to power outages from the centralized grid. Hospitals, police departments, and water-pumping stations already use on-site backup generators. The New York City Police Department has installed a fuel cell in a substation in Central Park because the cell costs far less than digging up the park to install underground power lines.[80]

Homeowners are also worried about disruptions caused by power outages. The first residential fuel cell was installed in a ranch-style house near Albany, New York, in June 1998.[81] A fuel-cell unit the size of a refrigerator can supply up to fifty kilowatts of electricity for powering a home.[82] In the future, distributed generation could become a boom business in the residential market. Millions of people now work out of their homes and rely on an uninterrupted flow of electricity to keep them connected over the Net. For them, backup generation might become a necessary cost of doing business. Many other homeowners have simply grown tired of power outages and a loss of heat, air conditioning, refrigeration, and other services and might opt to play it safe by installing their own micro-power plants. Elderly people that are homebound and dependent on life-sustaining medical equipment and services of all kinds might also install backup generators because they no longer trust that their power company will supply them with uninterrupted service. Last, a more security-conscious public, worried about crime and now terrorism, has expanded its security horizon and could soon include backup generation of electricity on its list of priorities. All in all, more than a million homeowners a year are already buying backup power systems of one kind or another. As fuel cells become cheaper and more convenient to install and use, they will likely become the most popular option.[83]

Distributed generation is also gaining favor because of the growing concern over global warming and the desire to use energy more efficiently in order to reduce CO_2 emissions. By locating mini-power plants at their homes or places of business, end users can exploit the heat generated by the electricity to heat the structure or to generate additional power. Co-generation greatly increases efficiency by reducing the amount of fuel used by as much as 50 percent. Co-generation also reduces CO_2 emissions by as much as 50 percent, because electricity and thermal energy do not have to be produced and transported separately to end users.[84]

A wholesale shift away from centralized power generation using fossil-fuel energy to hydrogen-powered fuel cells operating on a distributed generation grid—especially if the hydrogen is produced by using solar, wind, hydro, and geothermal forms of energy—could more dramatically reduce CO_2 emissions than could any other single development currently being pursued. The World Energy Assessment published jointly by the United Nations Development Program (UNDP), the United Nations Department of Economic and Social Affairs (UNDESA), and the World Energy Council (WEC) concluded that a near-zero-emissions-producing hydrogen-energy regime "would provide society with the capacity to achieve, in the longer term, deep reductions in CO_2 emissions . . . and thereby make it possible to limit CO_2 levels in the atmosphere to twice the pre-industrial level or less in response to climate change concerns."[85]

A growing number of customers are using distributed generation to engage in what industry people call "peak shaving." Power costs can vary from hour to hour in response to demand and generation availability. These fluctuations are converted into seasonal and daily time-of-use rate categories such as off-peak, on-peak, and shoulder rates. At peak periods, when demand is high, utilities often have to bring on line their older, less efficient plants. The additional cost is passed on to consumers in the form of spikes in electricity prices. At peak periods, owners of distributed-generation power plants can chose to go off the main grid and onto their own power supply to save money.

In the future, fuel cells will monitor pricing over an Internet connection or by way of digital signals embedded in the electricity itself, writes Peter Fairley in *Technology Review*. The fuel cells will analyze incoming information on, for example, the real-time price of natural gas and electricity, and, if the cost of switching over to distributed generation is favorable, the micro-power unit will automatically switch on. Since most commercial enterprises and homeowners are not experts in the complexities of the energy business, new intermediaries like Williams International, a Tulsa energy company whose pipelines already transport 20 percent of the U.S. natural-gas supply, will act as full-service providers. Williams International "provides a complete energy service package: financing the micropower unit, providing power from the grid, and helping consumers determine when peak shaving makes sense."[86]

Utility companies, interestingly enough, also serve to gain from distributed generation, although, until recently, many have fought the development. Because distributed generation is targeted to the very specific energy requirements of the end user, it is less costly and a more efficient way of providing additional power than is relying on a centralized power source. It costs a utility company between $365 and $1,100 per kilowatt-hour to install a six-mile power line to a three-megawatt customer. A distributed-generation system can meet the same electricity requirements at a cost of between $400 and $500 per kilowatt-hour.[87] Generating the electricity at or near the end user's location also reduces the amount of energy used because between 5 and 8 percent of the energy transported over long-distance lines is lost in the transmission.[88]

U.S. power companies are anxious to avoid making large financial investments in capital expansion because, under the new utility restructuring laws, they can no longer pass on the costs of new-capacity investment to their customers. And, because the field is now very competitive, power companies are reluctant to take funds from their reserves to finance new capacities. The result is that they put stress on existing plants beyond their ability to keep up with demand, leading to more frequent breakdowns and power outages. That is why a number of power companies are looking to distributed generation as a way to meet the growing commercial and consumer demands for electricity while limiting their financial exposure. The critical factor, from the utility companies' perspective, is controlling the distributed generation and letting it work "for rather than against them."[89]

Ann Chambers, in her book *Distributed Generation*, points to two strategies that power companies are using to "control the asset." The utility company could build and install new distributed-generation plants on its own transmission systems. It could also lease fuel cells to end users or put its own micro-power plants on the local sites and enter into an agreement with the user that, during peak-load periods, the user would go off the main power grid and use his or her own micro-power to ease the load and prevent massive outages across the entire system.[90] The end user would be compensated with an energy-bill discount.

The benefits of distributed generation, both for the power companies and the end users, are impressive. Arthur D. Little, the research and consulting firm, in its own lengthy review of the merits of distributed generation in 1999, concluded that "DG has the potential to play a major role as a complement or alternative to the electric power grid . . . the range of DG technologies and the variability in their size, performance and suitable applications suggests that DG could provide power-supply solutions in many different industrial, commercial and residential settings across the United States."[91] Even conservative industry analysts predict that distributed generation will produce up to 30 percent of all the new generating capacity in the U.S. in the future.[92]

The Hydrogen Energy Web (HEW)

The really great economic revolutions in history occur when new communications technologies fuse with new energy regimes to create a wholly new economic paradigm. The introduction of the printing press in the 1400s, for example, established a new form of communication that, when later combined with coal and steam technology, gave birth to the Industrial Revolution. Print provided a form of communication that was agile and quick enough to coordinate a world propelled by steam power. It would not have been possible to rely on script or oral communication to coordinate the increase in speed, pace, flow, density, and interactivity of commercial and social life made possible by steam power. Similarly, the telegraph, and later the telephone, provided forms of communication that were fast enough to accommodate the quickened pace, flow, density, and interactivity made possible when coal steadily gave way to an even more agile hydrocarbon, crude oil.

Today, hydrogen and the new fuel-cell distributed-generation technology are beginning to fuse with the computer and telecommunications revolution to create a wholly new economic era. The individual fuel cells that make up the growing distributed-generation revolution are just now being connected to one another with the help of sophisticated computer software, smart digital technologies, and Internet access to form the beginnings of a distributed-energy web. Soon, end users will not only produce their own electricity but be able to share it with others, posing a fundamental challenge to the current top-down, uni-directional energy regime currently in place around the world. "The transformation of passive energy users to freelance energy producers," writes Steve Silberman in *Wired* magazine, is "paralleling development in interactive media, peer-to-peer file sharing, and self governance" on the World Wide Web.[93] The consequences of connecting every owner of a fuel-cell micro-power plant with every other owner in an energy-sharing network will be as profound and far-reaching as was the development of the World Wide Web in the 1990s.

Noting the many striking similarities between what has already occurred with the World Wide Web and what is now getting under way with distributed generation, the Electric Power Research Institute (EPRI) concluded in its recent "Perspectives on the Future" that DG is going to unfold

> in much the same way the computer industry has evolved. Large mainframe computers have given way to small, geographically dispersed desktop and laptop machines that are interconnected into fully integrated, extremely flexible networks. In our industry, central-station plants will continue to play an important role, of course. But, we're increasingly going to need smaller, cleaner, widely distributed generators . . . all supported by energy storage technologies. A basic requirement for such a system will be advanced electronic controls: these will be absolutely essential for handling the tremendous traffic of information and power that such a complicated interconnection will bring.[94]

Many of the same considerations and concerns that led to the development of the World Wide Web are at work in the fledging development of the HEW. The Pentagon created the precursor to the Internet in the late 1960s. The Department of Defense (DOD) was anxious to save money on the cost of providing expensive new supercomputers to academic and defense contracting researchers, and it began to explore ways of sharing computers among people who were separated over long distances. The military was also concerned about the potential vulnerability to attack, or other forms of disruption, of centrally controlled communications operations. They were looking for a new kind of decentralized communications medium in which all of the parties could produce information and send it to one another in a way that would continue to function even if part of the system was disrupted or destroyed. The solution came in the form of the ARPANET, developed by the DOD's Advanced Research Projects Agency. The first host computer became operational in 1969. By 1988, more than 60,000 host computers were connected. The National Science Foundation soon created its own NSF net to connect university researchers across the country. When ARPANET shut down in 1990, the NSF net became the main vehicle for connecting computers and eventually metamorphosed into the Internet.[95]

In Denver, Colorado, two companies, Encorp and Celerity, are connecting five large distributed generators at commercial and industrial sites into one of the first energy micro-webs. Together, they can produce up to five megawatts of power. A similar energy micro-web is being put together by Celerity and a firm called Sixth Dimension in Albuquerque, New Mexico, to connect twelve generators. The Web will generate twenty-five megawatts of power.[96] In the future, "virtual power plants" could link thousands of fuel cells to generate as much electricity as 1,000-megawatt centralized power plants do today.

Already, utilities in thirty states allow customers to generate their own power and sell it back to the main grid.[97] And, in 2001, a bill was introduced in the U.S. Senate that would require all power plants to allow customers with on-site generators that run on renewable resources to sell power back to the main grid.[98]

Most DG power plants are still used as backups to the main grid and are only turned on in emergencies when power has been disrupted. This means that they remain idle most of the time. If they could be effectively integrated into the main grid, they could become a producing asset and supply incremental power during peak-load periods to power companies whose own capacities are too stretched to meet growing demand.

Eventually, the end users' combined generating power via the energy web will exceed the power generated by the utility companies at their own central plants. When that happens, it will constitute a revolution in the way energy is produced and distributed. Once the customer, the end user, becomes the producer and supplier of energy, power companies around the world will be forced to redefine their mission if they are to survive. A few power companies are already beginning to explore a new role as bundler of energy services and coordinator of energy activity on the energy web that is forming. In the new scheme of things, power companies would become "virtual utilities," assisting end users by connecting them with one another and helping them share their energy surplus profitably and efficiently.

Coordinating content rather than producing it becomes the mantra for power companies in the era of distributed generation, and America Online (AOL) becomes the business model to emulate.

But, before the HEW can be fully actualized, changes in the existing power grid will have to be made to assure both easy access to the Web and a smooth flow of energy services over the Web. Connecting thousands and then millions of fuel cells to main grids will require sophisticated dispatch and control mechanisms to route energy traffic during peak and non-peak periods. Encorp has already developed a software program for remote monitoring and control that would automatically switch local generators onto the main grid during peak loads when more auxiliary energy was required. Retrofitted existing systems are estimated to run about $100 per kilowatt-hour, which is still less costly than building new capacity.[99]

The problem with the existing power grid is that it was designed to ensure a one-way flow of energy from a central source to all of the end users. It is no wonder that Kurt Yeager, the president of the EPRI, recently remarked that "the current power infrastructure is as incompatible with the future as horse trails were to automobiles."[100] In many ways, the current grid resembles the state of the broadcast industry before the advent of the World Wide Web, when connections flowed only in one direction, from the media source to the viewing audience.

Transforming the power grid into an interactive network of thousands and then millions of small suppliers and users is a challenging task. Today's transmission systems are not set up to direct specific quantities of energy to specific parts of the grid. The result is that power flows all over the place, often causing congestion and energy loss. A new technology developed by the EPRI called FACTS (flexible alternative current transmission system) gives transmission companies the capacity to "deliver measured quantities of power to specified areas of the grid."[101] Silberman, writing in *Wired* magazine, makes an apt analogy with the World Wide Web, saying that we should "think of FACTS controllers as routers of the energy web."[102] American Electric Power (AEP) purchased the first FACTS in 1998 for its Kentucky operations, and nine utilities are now using it.[103]

The integration of state-of-the-art computer hardware and software transforms the centralized grid into a fully interactive intelligent energy network. Sensors and intelligent agents embedded throughout the system can provide up-to-the-moment information on energy conditions, allowing current to flow exactly where and when it is needed and at the cheapest price. Sage Systems, for example, has created a software program that allows utilities to "shed load instantly" if the system is at peak and stressed to the limit, by "setting back a few thousand customers' thermostats by two degrees . . . [with] a single command over the Internet."[104] Another new product, Aladyn, allows users to monitor and make changes in the energy used by home appliances, lights, and air conditioning, all from a Web browser.

In the very near future, sensors attached to every appliance or machine powered by electricity—refrigerators, air conditioners, washing machines, security alarms—will provide up-to-the-minute information on energy prices, as well as on temperature, light, and other environmental conditions, so that factories, offices, homes, neighborhoods, and whole communities can continuously and automatically adjust their energy requirements to one another's needs and to the energy load flowing through the system.

A study done for the U.S. Department of Energy and published in 2000, entitled "Making Connections," concluded that the biggest obstacle to creating an interactive energy web is "the long-standing regulatory policies and incentives designed to support monopoly supply and average system costs for all ratepayers." The government-commissioned study made clear that, in the existing regulatory environment,

> utilities have little or no incentive to encourage distributed power. To the contrary, regulatory incentives drive the distribution utility to defend the monopoly against market entry by distributed power technologies.[105]

In the two years since that report was issued, the situation has begun to change. As mentioned earlier, a growing number of utility companies are coming to the conclusion that the advantages of encouraging distributed generation outweigh the advantages of maintaining their monopoly control over the grid and are beginning to work with independent owners and operators of on-site distributed-generation power plants to create interactive energy webs.

Currently, one of the pressing needs is to establish uniform standards to assure owners of distributed-generation equal access to the grid. The U.S. Department of Energy, the Institute of Electrical and Electronics Engineers, power companies, independent producers of fuel cells and other distributed-generation technologies, and end users are currently wrestling with regulatory barriers and anticompetitive practices of some power companies that prevent distributed-generation systems from securing fair and equal access to the grid.

The coming together of distributed-generation and distributed-system intelligence changes the energy equation forever. For the first time, the potential exists to replace a traditional top-down with a new bottom-up approach to energy—a democratization of energy, in which everyone can be his or her own vendor as well as consumer.

Distributed generation, and the creation of a regional energy web and, eventually, a worldwide one are the logical follow-ups to the creation of the worldwide communications web. Interactive communications and interactive energy-sharing each compliments and feeds off the other. As the two technology revolutions continue to fuse, the foundation is laid for a new type of economy and society—one in which, at least in theory, the increase in energy flow-through can be met with a new kind of complexity that is, for the first time in history, decentralized in nature and truly democratic in form.

Turning the Car into a Power Plant

The distributed-generation revolution is likely to take off in the next few years, with the introduction of automobiles, trucks, and buses operated by fuel cells. Every major automaker in the world has announced plans to introduce fuel-cell-powered automobiles. In 1997, Daimler-Benz launched a $350 million joint effort with Ballard Power Systems, a Canadian firm and leader in fuel-cell development, to create hydrogen fuel-cell engines. The automaker says it will produce 100,000 fuel-cell cars by the end of the decade—one-seventh of its total current production. Ford subsequently joined with Daimler-Chrysler and Ballard Power Systems, upping the joint investment to more than $1 billion.[106] Toyota hopes to have fuel-cell cars on the road within the decade. GM has promised to have fuel-cell cars ready by 2010. Nissan, Honda, and Mitsubishi have also announced plans to produce hydrogen-powered fuel-cell cars and have between them committed another $1 billion to the effort.[107]

Although the public has heard little or nothing of hydrogen-fueled automobiles, behind the scenes the world's automakers are gearing up for what amounts to the most important revolution in how power is harnessed since the introduction of the internal combustion engine 100 years ago. Bill Ford, the great-grandson of Henry Ford and the current chairman of the Ford Motor Company, has gone so far as to say, "I believe fuel cells will finally end the 100-year reign of the internal combustion engine."[108]

The enthusiasm of the automakers has been matched by at least a few of the world's leading energy companies. Chris Fay, the chief executive of Shell U.K., London, has said that at Shell "we believe that hydrogen-fuel-cell-powered cars are likely to make a major entrance into the vehicle market throughout Europe and the U.S. by 2005." Fay says that "this trend poses a real challenge to a company like Shell to develop new products and new technologies and to prepare and inform our customers for the changes that lie ahead."[109]

The full implications of this shift are momentous. There are 750 million passenger and other vehicles on the world's roads, and that number is expected to double over the next twenty-five years.[110] They are powered by fossil fuels.[111] In the U.S. alone, transportation accounts for 54 percent of all the oil consumed each year.[112] Transportation accounts for more than 20 percent of global primary energy.[113] Moreover, according to the International Energy Agency, 17 percent of the world's carbon dioxide emissions come from the burning of oil in road transportation.[114]

Now, say industry observers, imagine if the entire global fleet of automobiles, buses, and trucks were to be powered by hydrogen fuel cells rather than the internal combustion engine. Of course, at first the hydrogen would have to be produced by steam-reforming methanol—unfortunately, some automobile companies even want to use gasoline. But, eventually, renewable resources—photovoltaic, wind, hydroelectric, geothermal, and biomass—will be used to electrolyze water cheaply and efficiently, separating out hydrogen to be used directly as a fuel, thus bypassing the fossil fuels as a hydrogen source. Hydrogen fuel cells have zero emissions. As mentioned earlier, the only by-products are heat and pure water. The long reign of hydrocarbon energy would come to a close and, with it, the increasing entropy bill in the form of CO_2 emissions from the burning of carbon fuels. Global warming could be drastically slowed to only twice the pre-industrial level, and the long-term environmental risk of heating up the planet could be substantially mitigated.

Equally important, in the new hydrogen-fuel-cell era, the automobile itself is "a power plant on wheels" with a generating capacity of twenty kilowatts.[115] Since the average car is parked about 96 percent of the time, it can be plugged in, during non-use hours, to the home, office, or main interactive electricity network, providing premium electricity back to the grid. The revenue gained from selling energy back to the grid could help defray the cost of the lease or purchase price of the vehicle. If just a small percentage of drivers used their vehicles as power plants to sell energy back to the grid, most of the power plants in the country would be eliminated altogether. That is because a hydrogen-fuel-cell-powered transportation fleet of 200 million vehicles has four times the generating capacity of the entire national power grid.[116] The potential energy made available by switching over to hydrogen-fuel-cell automobiles is incredible. Bertrand Dusseiller of Asea Brown Boveri calculates that the rated prime-mover power of all the automobiles manufactured each year is greater than the total rated capacity of all the world's power stations.[117]

The key question facing the automobile industry during the transition to hydrogen-fuel-cell-powered vehicles is how to produce, distribute, and store hydrogen cheaply enough to be competitive with gasoline at the pump. Some studies estimate that it would cost more than $100 billion to create a national infrastructure for producing and distributing hydrogen in bulk. The "hydrogen question" is the classic chicken-and-egg problem. The automobile companies are reluctant to manufacture direct-hydrogen fuel-cell cars for fear that the energy companies won't invest sufficient funds to create thousands of hydrogen refueling stations. That is why the car companies are hedging their bets by developing fuel-cell cars with on-board reformers that can convert gasoline and natural gas to hydrogen. The energy companies, in turn, are nervous about committing billions of dollars to create a national infrastructure to support hydrogen refueling stations if not enough direct-hydrogen fuel-cell vehicles are manufactured and sold.

Critics argue that if the automobile industry commits to fuel-cell cars with an on-board fuel processor—a portable thermochemical plant—to convert gasoline or methanol into hydrogen, it will be locking itself into an expensive and unnecessary long-term strategy that could cost more than $1 trillion for the next car fleet.[118]

Writing in the *International Journal of Hydrogen Energy*, C.E. "Sandy" Thomas, formerly of Directed Technologies Inc.—a firm that advises the Ford Motor Company on fuel-cell technology—and his colleagues argue that, on the basis of an exhaustive study of the comparative cost of using on-board fuel processors to convert gasoline or methanol to hydrogen versus fueling the vehicle with direct hydrogen, the latter is actually less costly. The authors contend that the reason hydrogen costs are so high in all of the calculations done by industry is that the studies assume that a nationwide pipeline system for transporting hydrogen, similar to natural-gas pipelines, would have to be built at a cost of tens of billions of dollars. However, in a study conducted under contract for the Ford Motor Company and the U.S. Department of Energy, the authors conclude that "hydrogen can be delivered to the fuel-cell vehicle at lower costs by producing and installing small-scale steam methane reformers or small-scale electrolyzers at the local fueling station or fleet operator's garage," thus avoiding the need to build a massive hydrogen pipeline system.[119] In the transitional period, hydrogen would be produced "where and when it is needed in quantities that match the incremental growth of fuel-cell vehicle (FCV) sales, minimizing the need for multi-billion-dollar investments prior to the introduction of sufficient numbers of FCVs to provide adequate return on investment."[120] Moreover, the conversion process would rely on using existing natural-gas pipelines to provide natural-gas feedstocks to produce the hydrogen or, alternatively, water electrolysis connected to the electricity grid to produce hydrogen.

When the researchers compared the methanol and gasoline on-board reforming process to the direct use of hydrogen drawn from small-scale steam methane reformers and electrolyzers located at existing fueling stations, they found that direct hydrogen is the less-costly fuel. A direct-hydrogen vehicle would also be less costly than vehicles with on-board methanol or gasoline reformers. According to the study, a methanol FCV would cost between $550 and $1,600 more than a direct-hydrogen fuel-cell vehicle, and the gasoline FCV would cost between $1,600 and $4,500 more.[121]

Finally, there is the issue of whether hydrogen can be feasibly stored on board vehicles. An earlier study conducted for the California Air Resources Board concluded that compressed hydrogen takes up more volume than gasoline and methanol tanks and would take up so much passenger and trunk space that the vehicle would be unworkable. Sandy Thomas, who is currently president of H_2 Gen, counters that the storage issue has been resolved at Ford. The company has redesigned the vehicle in such a way that the hydrogen tank fits very well and provides a range of 380 miles.[122]

The Hydrogen Technical Advisory Panel to the U.S. Department of Energy, in a 1999 report, concurred with the assessment made by Thomas and his colleagues in the *International Journal of Hydrogen Energy*. On the question of assuring adequate on-board storage capacity for direct hydrogen, the panel concluded that "compressed gas or liquid in sufficient quantities for acceptable vehicle range between refueling, and without loss of passenger or cargo space, requires no breakthroughs in hydrogen storage technology."[123]

The Department of Energy Advisory Panel also argued that "on a cost-per-vehicle basis—which includes the vehicle cost and a portion of the fuel-supply infrastructure allocated to each vehicle—the costs of direct hydrogen and on-board fuel processor fuel-cell vehicles can be comparable after the vehicle population becomes large enough to amortize the cost of the refueling rate."[124]

The panel expressed concern that if the automobile industry pursued a phase-in strategy using on-board processors to convert gasoline or methanol to hydrogen in the hope of selling sufficient cars to economically justify a transition to direct-hydrogen fuel-cell vehicles in a decade or so, "it could lock out direct-hydrogen vehicles for many decades to come, thus denying society the superior long-term benefits of direct-hydrogen vehicles." Those benefits include zero emission of CO_2, and reduced dependence on foreign oil.[125] The DOE panel also agreed with the study by Thomas and his colleagues that, in the early stages, small on-site hydrogen generators, either steam reformers connected to the nation's gas pipelines or electrolyzers connected to the nation's electrical grid, could be installed at local refueling stations to produce and store hydrogen. The panel suggested that, as the number of fuel-cell vehicles grew, large central plants, each delivering hydrogen to fueling stations by hydrogen gas pipelines, could be established, or on-site production of hydrogen using small generators could continue to serve consumer demand.[126]

Germany became the first European country to open a commercial hydrogen refueling station in Hamburg on January 13, 1999. At the opening ceremony, the city's mayor, Ortwin Runde, asked his fellow citizens to try to imagine what a hydrogen-fueled transport would mean for the quality of life in cities like Hamburg. He mused:

> The streets will be quiet. Only the sound of tires and rushing wind will accompany passing vehicles instead of the roar from exhaust pipes. The city will be clean, since emissions will be practically zero. Pedestrians strolling on the sidewalks won't be turning up their noses, guests won't be fleeing from the streets' stench into the cafés because now they can enjoy the sundowners in the open air.[127]

Amory B. Lovins and Brett D. Williams of the Rocky Mountain Institute have suggested a somewhat different approach for securing direct hydrogen for fuel cells. Their plan relies on using stationary fuel-cell power plants in homes and offices to provide the hydrogen energy in the early stages of the transition to a hydrogen fleet. Lovins and Williams point out that more and more buildings are expected to install on-site fuel-cell power plants in the next few years. Their proposal calls for using fuel-cell appliances in homes and offices during spare off-peak capacity periods to produce hydrogen: "You park your fuel cell Hypercar at work (or at your house or apartment . . .), you plug into both the electricity grid and a snap-on fuel line bringing surplus hydrogen from the fuel appliance in the building."[128]

There is one more issue that ought to be raised in regard to hydrogen fuel: the public perception that it may be dangerous. Much of the public's angst about hydrogen goes back to an event that occurred in 1937. The German airship *Hindenberg* caught fire while attempting a landing in Lakehurst, New Jersey, killing thirty-six of the passengers aboard. Contrary to public belief, the *Hindenberg* did not explode, and hydrogen was not the cause of the fire. In 1997, Addison Bain, the former head of the hydrogen program at the Kennedy Space Center, presented the findings of a decade-long study of the accident. According to Bain's research, the likely cause of the fire was static electricity in the air that provided the spark that set fire to the cotton fabric covering the ship. As the fire spread, it ignited the hydrogen—but the hydrogen was not the initial cause of the fire.[129]

Studies over the years have concluded that hydrogen is no more dangerous than other fuels. It may even be safer in some situations, because it quickly evaporates when released instead of spreading along the ground as does gasoline. A study commissioned by the German *Bundestag* in 1993 and carried out by Germany's Office for the Assessment of Technology Consequences concluded that "the technical risks in all components of a hydrogen energy system, from production to utilization are, in principle, regarded as controllable."[130] Currently, international standards for the safe production, storage, transport, and use of hydrogen are being worked out by the International Organization for Standardization in Geneva, Switzerland.

Although there is likely to be a great deal of discussion and wrangling and more than a few mis-starts along the way, the consensus is that this coming decade will see the beginning of the end for the internal combustion engine and the dawn of the hydrogen-propelled fuel-cell car. At the 2002 North American International Auto Show in Detroit in January 2002, General Motors wowed the crowd with its new prototype hydrogen fuel-cell car, the "Autonomy." The car is a revolution in automobile design. The sleek space-age body is snapped together and modular, allowing the user to change body types and styles at will; the chassis is designed to last for more than twenty years. The car itself is run by software and eliminates the mechanical systems found in conventional automobiles, including the engine; fixed steering column; pedals for the brakes, clutch, and accelerator; and the gearshift lever. These functions are now managed by computer software operated by the driver using a single control stalk. GM says that, because there are so few components, the new hydrogen-powered fuel-cell vehicles will eventually be cheaper to buy and safer to drive than those with internal combustion engines. In the fall of 2002, GM debuted a more advanced version of its fuel-cell car, the Hy-wire, at the Paris Motor Show and vowed to have a mass-produced hydrogen-powered fuel-cell automobile in the showrooms by 2010. The company is currently spending more than $100 million per year to develop a production version of the Hy-wire, and company CEO Richard Wagoner has said repeatedly that the automaker's goal is to be the first car company to produce one million fuel-cell vehicles.[131]

The same week that General Motors introduced its prototype hydrogen-powered fuel-cell automobile, the U.S. Department of Energy, in a dramatic turnaround, announced that it was abandoning a $1.5 billion eight-year joint project with GM, Ford, and Daimler-Chrysler to develop high-mileage gasoline-fueled vehicles and instead was committing itself to help develop hydrogen-powered fuel-cell vehicles. According to Secretary of Energy Spencer Abraham, the new Department of Energy program, entitled "Freedom Car," is "rooted in President Bush's call, issued last May in our National Energy Plan, to reduce American reliance on foreign oil."[132]

Then, in April 2002, Governor John Engler of Michigan unveiled a long-term economic-development plan that would make the state the world leader in developing, manufacturing, and marketing fuel cells and related hydrogen-based technologies, goods, and services. He said that his goal was to work with the automakers to ensure that Detroit and the state of Michigan maintain their preeminent status as the automotive capital of the world. The Michigan plan, called "NextEnergy," includes the establishment of a NextEnergy Center that will serve as a clearinghouse for hydrogen-energy technologies, and the creation of the NextEnergyZone, a 700-acre state-owned campus that will be the high-tech center for hydrogen innovation. The campus will be designated a tax-free Renaissance Zone in the hope of luring companies from around the world to set up their hydrogen research-and-development centers there. Companies will also receive tax rebates for jobs that they create in the zone.[133]

The plan calls for close collaboration with the state's educational system to design curricula and prepare the next generation with the requisite technical skills to pursue cutting-edge research and development and to manufacture and market hydrogen technologies, products, and services.

The Michigan plan also calls for the creation of a National Alternative Energy Program funded, in part, by the federal government. The program would serve as an underwriter's laboratory for developing industry standards and certification systems.

To encourage the widespread dissemination of stationary and vehicular fuel-cell technologies and to hurry the transition to a hydrogen economy, governor Engler announced the state's intention to exempt from the sales-and-use tax individuals and institutions who purchase these technologies. He said that the state would also work with the utility industry to create hydrogen-energy microgrids to demonstrate the feasibility of hydrogen power.

The governor likened the ambitious effort to the creation of Silicon Valley, which spawned the software revolution in the last quarter of the 20th century.

Noting that the hydrogen economy would grow to about $100 billion by 2010 and produce hundreds of thousands of new jobs, Governor Engler said that Michigan hoped to be the global center for the transformation to a hydrogen era.

Other states, including Ohio and California, are launching their own initiatives to capture the emerging hydrogen energy market.[134]

The European Union Leaps Ahead

The European Union (EU) caught the world by surprise in October 2002 with the dramatic announcement by Romano Prodi, the president of the European Commission, the governing body of the EU, that Europe intended to become the first fully integrated renewable-based hydrogen superpower of the 21st century. (The author serves as a personal adviser to President Prodi and, in that capacity, prepared the strategic white paper that led to the EU hydrogen-energy initiative.)

The EU had already committed itself to making the transition from fossil-fuel dependency to a renewable energy future. The EU's renewable energy targets are the most ambitious in the world. By 2010, 22 percent of the electricity and 12 percent of all the energy produced in the EU has to be derived from renewable energy sources.[135] The EU realized that the kind of renewable energy future it envisioned would be impossible without using hydrogen as a means for energy storage.

To make a renewable-based hydrogen energy future a reality, the EU has forged an unprecedented public-private partnership with the European business community and civil society. Together, the three sectors have established an operational road map for making a phased transition to a hydrogen-energy regime and have established working groups to foster research, development, and commercial introduction of the new fuel-cell technologies. The EU also greatly expanded its financial commitment from £127 million for sustainable energy development over the past four years to £2.1 billion for 2003 through 2006.[136]

In announcing the new energy plan, President Prodi remarked that the transition to a hydrogen era and decentralized distribution of energy across Europe signals the next major step in the integration of Europe after the successful introduction of the Euro. He likened the Herculean effort to the American space program of the 1960s and 1970s that put a man on the moon and helped spawn the high-tech economy of the 1980s and 1990s.[137] President Prodi is mindful that England became the leading economic power in the world in the 19th century because it was the first country to use its vast reserves of coal to power the steam engine. Similarly, the U.S. became the world's leading economic power in the 20th century by using its vast oil reserves to power the internal-combustion engine. Mr. Prodi envisions the European Union becoming the world's leading superpower in the 21st century by using hydrogen to power fuel cells and believes that Europe's future success depends, in no small measure, on its ability to phase in the new energy era in a timely fashion.

Europe's hydrogen initiative acted as a stimulus to U.S. companies' efforts to create a public-private partnership in America. A coalition of U.S. companies lobbied the Bush administration in the fall of 2002, seeking a federal government commitment of $5.5 billion over the next ten years for research and development of hydrogen fuel, technology, and infrastructure.[138] In his January 2003 State of the Union address, President Bush said that his administration was committed to the pursuit of a hydrogen future and would introduce a series of proposals in the new Congress to move the U.S. to the front in the hydrogen race.[139] The subsequent energy bill presented to Congress by the White House and the Republican Party, however, was quite different in its orientation to the agenda set forth in the European Union. The Republican energy budget proposal of $1.7 billion over the next five years calls for expanded funds to the coal, natural gas, and nuclear industry to research and develop hydrogen extraction using these traditional energy resources, while earmarking little additional funds for the research and development of renewable energy to extract hydrogen.[140] Many observers—especially within the environmental community—complained that the Bush administration was using hydrogen as a Trojan horse to bolster the interests of his friends in the fossil-fuel and nuclear industries, at the expense of pursuing a renewable-based hydrogen future. It appeared, at least to those individuals and organizations working in the environmental field, that the White House had set its sights on advancing a hydrogen future without moving beyond a fossil-fuel past.

The European Union, by contrast, while it sees fossil fuels as an important part of the energy mix for the next several decades, is engaged in what it calls a parallel track strategy. On the first track, the EU is committed to conserving existing fossil-fuel energy in the years immediately ahead, with tough automobile fuel-efficiency standards and other conservation practices, and an unswerving commitment to the targets set out in the Kyoto Protocol on global climate change. On the second track, the EU is committed to underwriting and encouraging the speedy development of renewable technologies and a hydrogen-energy infrastructure so that Europe can phase out its reliance on fossil fuels, over time, while it lays the foundation for a new energy era.

THE HYDROGEN ECONOMY is within sight. How fast we get there will depend on how committed we are to weaning ourselves off of oil and the other fossil fuels. If we simply toy with or delay the transition in the belief that there is plenty of cheap oil left to supply our needs well into the middle years of the 21st century, we may find ourselves wholly unprepared to make a timely transition were global oil production to peak in the next few years. Building a new infrastructure to support a mature hydrogen economy will be complicated and expensive. That infrastructure could be laid down in less than a decade if the enthusiasm and commercial zeal were there. After all, the infrastructure for the Internet economy and the World Wide Web was put in place, at least in the developed world, in less than a decade, fundamentally changing the way that companies do business and that people communicate with one another. Many of the leading business magazines are predicting that the hydrogen economy and the worldwide energy web will be the next great commercial revolution. Turning these forecasts into reality, however, requires both commercial and public commitment to a hydrogen future and a practical vision on how to get there.

For the first time in human history, we have within our grasp a ubiquitous form of energy, what proponents call the "forever fuel." Hydrogen will eventually be as cheap as personal computers, cell phones, and palm pilots. When that happens, the possibility opens up to truly democratize energy, making it available to every human being on Earth.

9

REGLOBALIZATION FROM THE BOTTOM UP

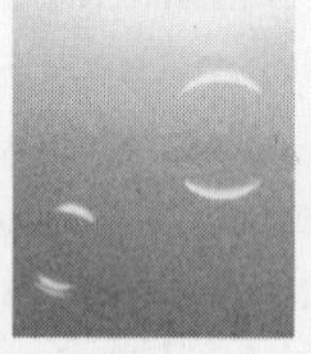

Our future lies with hydrogen. But who will control the "forever fuel"? Hydrogen, because of its universality, offers the prospect that we might be able, at long last, to democratize energy and empower every human being on Earth. But, while the opportunity exists, there is no guarantee that hydrogen will, in fact, be fairly and equitably shared among all people. Much depends on how we come to "value" hydrogen. Will we see it as a shared resource, like the sun's rays and the air we breathe, or as a commodity, bought and sold in the open marketplace, or something in between?

Lessons from the World Wide Web

Determining the "status" of hydrogen will set the ultimate course for the future of the hydrogen economy and will have deep consequences for the political and social institutions that will grow up alongside the new energy regime. The question of hydrogen's status bears many parallels with the question of the status of "information" on the Internet. Early advocates of the Internet and the World Wide Web argued passionately for the idea that "information ought to run free." The very architecture of the new communications medium, they contended, favored a free sharing of information between people. The World Wide Web, after all, belongs to no one and is open to everyone. It is merely every computer user in the world being connected to any other for the purpose of entering into a conversation. "The economy of the future," exclaimed cyberspace theorist John Perry Barlow, "will be based on relationships rather than possessions."[1] Conversations and the exchange of information between people ought not to be held at bay or be made subject to access fees, licenses, and permission of various kinds, goes the argument. The purists say why not download music, make copies, and share it over the Net with others for free? Why not make copies of articles and add your own thoughts to stories and reports and send them along to friends, and even strangers, for free? The question of what ought to be regarded as free information and what should be subject to payment to copyright holders or to gatekeepers has been the central debate as this new global interactive communications medium has come of age.

Those who would keep information running as free as possible on the Net stake their claims on several of the key operating assumptions of the Web. First, the Web is designed so that every user can become a supplier of content. The Web eliminates the old centralized hierarchy of mass broadcasting in which communication flowed one way from giant content providers to passive individual consumers. Second, because the Web is a network, anyone on it can engage in one-to-one as well as many-to-many interactivity. That means that literally anyone can have access to potentially everyone else on the Web. This kind of personal power to communicate is unprecedented in human history. Suddenly, everyone has access to everyone else, in a kind of instant democratization of communication. In a decentralized distributed-communications network where there is no central locus of control, no one, at least in theory, is in charge, and everyone is empowered. The idea of getting permission from some central authority or gatekeeper to communicate ideas, messages, and information of various kinds to others on the network undermines the very nature of what a network is all about. Here, all of a sudden, is a medium in which everyone can command his or her own center. Every node of the network becomes a staging area for sharing creativity. In short, the Net's architecture gives everyone far greater potential control over the information that they send and receive. It is easy to understand, then, how enthusiasts might be irked or even angered at the prospect of being denied the full freedom of expression by censors, copyright holders, or gatekeepers.

The great promise of the Net, however, has been compromised, at every step of the way, by commercial interests determined to gain a foothold over the medium. Companies like Microsoft and AOL Time Warner have been relentless in their efforts to tame the World Wide Web and change it from an open channel for the free flow of communication and information to a proprietary domain where logging on, accessing, downloading, and sending information are all paid-for affairs. When Napster and other upstarts began providing software so that millions of people could copy and share their music libraries with one another—peer-to-peer file sharing—free of charge, the major recording companies struck back with lawsuits, contending that because the music was copyrighted, payment for use was required. The music giants prevailed in court, and now Napster has reinvented itself as a commercial medium, making available access to the vast Bertelsman Napster music library for a monthly subscription fee, with a percentage of the fee going to the music companies and recording artists to satisfy copyright arrangements. Other renegade cyberspace players, however, continue to fight back. Arguing that music ought to run free, they are setting up new software programs to skirt the law and allow anyone and everyone to freely share music files.

The battle between, on the one hand, cyber-activists who advocate free access and the sharing of information, and, on the other hand, commercial interests who favor privatizing and commodifying cyberspace is being fought out across the medium. Nowhere are the stakes higher than in the sharing of code between users—the so-called open-source movement. For example, Linux is an operating system that is available for free. Moreover, the source code itself is made openly available to everyone to examine and modify. Its developers share their work with everyone else and encourage all the other users of the operating system around the world to work with one another collectively to solve problems and create new code. If a user of the Linux operating system has a question or a problem that needs resolution, he or she posts a query on a Linux-help Web page, and, usually within minutes, other users reply and help to find a workable solution. Linux boasts between 10 million and 20 million users and is the fastest-growing operating system in the world.[2] Some industry observers predict that Linux might, in time, even begin to challenge Microsoft's control of the operating-system business.

Although the battle to control content on the Net has tilted in favor of commercial interests in recent years, the struggle is far from over. The question of what is free and what is not will continue to dog the new medium. In distributed networks, in which everyone is a potential producer as well as a consumer of content, and in which each person has access to the whole of the network, which will very soon be one out of every six people on Earth, expect an escalating tug-of-war between those who argue that information ought to run free and commercial interests who want to exact some kind of toll.

Look for a similar struggle to unfold around the hydrogen energy web (HEW). Certainly, it can be argued that just as a great deal of the conversation and information that flows between people—but obviously not *all* communication and information—ought to run free, a like-minded case could be made that hydrogen should be considered a free or shared good, available to all. After all, it is the most basic and universal element in the whole universe.

The Hydrogen Commons

The issue of whether either human thought or the elemental energy of the universe ought to be considered a freely shared resource held in common or a propertied commodity goes to the heart of one of the deepest questions that the human family has had to grapple with over history: To whom does everything that constitutes life belong?

In the late medieval era, a great debate unfolded between the Church and an incipient merchant class over the question of whether time itself was a universal gift and therefore free, or something that could be owned and profited from by the charging of interest. The merchants argued that "time is money," and that charging interest was a legitimate form of compensation for allowing someone to use the lender's money for a period of time. The Church argued that usury was a mortal sin, but not just because it was exploitative in nature. Rather, Church authorities challenged the very legitimacy of the act itself. How could merchants profit from the selling of time, asked Church leaders, when time does not belong to them but rather to God, who gives it up freely as a gift to human beings so they can prepare for their salvation? Thomas Chobham wrote:

> The usurer sells nothing to the borrower that belongs to him. He sells only time which belongs to God. He can therefore not make a profit from selling someone else's property.[3]

In the 16th century in England, a second great debate emerged over what ought to lie in common and what could be claimed as private property. Land in medieval Europe was considered a divine trust, lent to human beings to till and cultivate. Although feudal landlords exercised proprietary rights over the land and leased it to peasant farmers under various tenancy arrangements, the land itself could not be easily divided, bought, or sold. Moreover, much of the land was held in common and managed collectively by peasant farmers. Beginning in the 1500s, during the Tudor reign in England, acts of Parliament were passed that allowed landlords to enclose commonly held land—that is, purchase it in the form of private property and, by so doing, sever any previous rights that tenants enjoyed in sharing the fields collectively. The great enclosure movements, which began in England and later spread to the Continent under various guises, ended a 600-year feudal period in which peasants belonged to the land. Henceforth, the great landed feudal estates would be divided up in the form of private real estate, owned by the local aristocracy and wealthy peasant farmers. Land, once held in common as a shared good, was reduced to private parcels of property that could be bought and sold in the open market.

In the 18th century, governments began to enclose the oceanic commons by claiming sovereignty over coastal waters extending three miles from shore—that was the distance that artillery could reach at the time.[4] A far more aggressive drive to enclose the oceanic commons began after World War II with an announcement by President Truman that the United States would extend its coastal waters to include "jurisdiction and control" of the gas and oil deposits and minerals along the seabed of the continental shelf. Truman's proclamation touched off a frenzy of counterclaims by other nations, each seeking to enclose "its" continental shelves and offshore fisheries. By the early 1970s, seventeen nations had claimed sovereignty over coastal waters extending 200 miles out to sea.[5]

A law-of-the-seas convention was drafted in 1982 under the auspices of the United Nations, guaranteeing signatory nations sovereignty twelve miles out to sea and exclusive economic rights 200 miles out on open oceans. These exclusive economic zones gave each nation "sovereign rights for the purpose of exploring and exploiting, conserving and managing the living and nonliving resources of the oceans, seabeds, and subsoil."[6] The great ocean grab effectively enclosed 36 percent of the world's ocean areas, containing 90 percent of the commercially exploitable fish and 87 percent of the projected offshore oil reserves along continental shelves. Although many nations have yet to ratify the convention, it has established the new parameters of state sovereignty, bringing much of the oceanic commons under the grip of enclosure.[7]

In the early 20th century, the atmospheric commons over countries were enclosed into air corridors that became the sovereign domains of governments. Governments, in turn, charge commercial airlines for the right to use these corridors. Governments have further chipped away at the atmospheric commons by allowing commercial enterprises to sell and buy "air rights" above their property. In the past few years, a debate has emerged in Washington over whether the radio frequencies that make up the electromagnetic spectrum, still a commons held in trust by governments and leased to broadcasters, should be sold off to commercial enterprises and be transformed into private "electronic real estate" that could be traded in the global marketplace.

Another equally passionate debate is emerging as to whether or not the global gene pool—the legacy of millions of years of biological evolution and long considered a commons—ought to be enclosed and made the private intellectual property of life-science companies. To date, the U.S. and most European nations have declared that genes and the proteins they code for, as well as living cells, tissues, organs, embryos, and whole species, are potentially patentable subject matter. Other nations and many of the world's leading nongovernmental organizations argue that the gene pool is, by nature, a "shared commons" and not reducible to either political or commercial property.

What about human thoughts? Should they be sold as a commodity, or freely shared as a commons? Before the modern era, the very idea that one could own thoughts, or charge someone else for hearing or using one's thoughts, would have been considered bizarre. In oral cultures, for example, when ideas and stories were passed on by word of mouth and embellished and altered with each successive telling, no one could claim absolute ownership. The introduction of printed material, however, made it easier to possess ideas and make proprietary claims over thoughts. Now, say the advocates, the new communications technologies—computers, software, and the World Wide Web—make communications more akin to oral culture because they allow for more of a genuine collective participation in the sharing of ideas and thoughts than was possible during the reign of print and the mass media.

What about the use of hydrogen distributed over the hydrogen energy web? Should we consider hydrogen a public good to be collectively shared, or a private resource subject to profit in the marketplace? Or some combination of both? Of course, if hydrogen existed free-standing everywhere in nature like the air we breathe and was therefore immediately available to do work for us without any cost, we would no doubt consider it a free good. But hydrogen does not come to us in an easily usable form. It must be extracted from something else in nature—from a fossil fuel, or biomass, or water—and then pumped into a fuel cell to perform electrical work.

So, while hydrogen exists everywhere, and is therefore not a scarce resource, human ingenuity has to pluck it from its surroundings and harness it to generate power. The extraction process requires an investment of time, labor, and capital, as do the storage and use. But, as the cost of producing hydrogen energy continues to go down—which it will—its status as a public good will continue to go up, because it is evenly dispersed around the world and unlimited in supply, unlike fossil fuels. It is possible to imagine a future, perhaps not more than 100 years from now, in which the cost of producing unlimited amounts of hydrogen is virtually zero. All earlier sources of nonrenewable fossil-fuel energy conform to a classic Hubbert bell curve. That is, they are expensive to process at first, then become cheaper as the enabling technologies become less costly and more sophisticated, and then eventually become more expensive to process as the reserves become more scarce. Hydrogen production, however, is a straight line projecting ever upward. That is, it becomes cheaper and cheaper to produce more of it. After a while, the only substantial cost is maintaining and improving the smart energy networks that the hydrogen flows through around the world.

If hydrogen is going to become less and less expensive to produce, and eventually will become an "almost" free resource, but if the smart networks that it runs on are going to be costly to build and maintain, then we need to think seriously, at the beginning of the hydrogen age, about the kind of institutional framework to put in place that best reflects the character of the energy source we are using. The HEW and the hydrogen economy built from it require a radical new kind of architectural design that brings private and public, profit and nonprofit ways of doing business into a symbiotic relationship that reflects both the proprietary and public aspects of the new energy regime.

Democratizing Energy

Early efforts to democratize the flow of information on the World Wide Web met with some limited success but, as mentioned, was quickly eclipsed by corporate gatekeepers like AOL. The community networking movement spread across the U.S. in the late 1980s and early 1990s to encourage greater participation in the life of the communities they served. Free-nets offered free online access as an inducement to become involved. The nets were noncommercial and had no advertising, nor did they charge subscription fees. The free-nets helped build public awareness of the Net and drew in a lot of the "early adopters," in large measure because access was free. Unfortunately, content on the community networks was often sparse and unentertaining, and it failed to keep the interest of online users. Meanwhile, companies like AOL began to fill the vacuum, providing cheap and easy access to the Net as well as more engaging and entertaining fare. Most of the hundreds of Internet-based community networks have been discontinued, although more than 100 are still operational and, in some communities, quite successful.[8]

Today, communities of interest rather than communities of geography have begun to flourish on the Web. Civil society organizations (CSOs), in particular, are using the Net as an organizing tool to unite like-minded interests in common pursuits in areas as diverse as agricultural reforms, animal rights, social justice, human rights, economic reforms, women's issues, public health, and the arts. The "democratization of information" is finding its place on the World Wide Web as communities of interest share ideas, tasks, and goals to advance social agendas and build social capital. Interestingly, the Web has become a powerful global forum for building communities even as it has fallen prey to commercial interests and been reduced, in part, to corporate dominance. The Web has, in effect, become a hybrid communications medium that is both commercial and social—

partially enclosed, commodified, and privatized while partially open to serve as a vast social commons for the free sharing of ideas and interests.

Distributed generation and the HEW are in the very early stages of development, much like the Internet was in the late 1980s. The way that distributed generation is structured during the takeoff stage in the next five years will likely determine the energy infrastructure that eventually evolves and matures ten to fifteen years from now.

The first thing to keep in mind is that with distributed generation, every family, business, neighborhood, and community in the world is potentially both a producer and a consumer and a vendor of its own hydrogen and electricity. Because fuel cells are located geographically at the sites where the hydrogen and electricity are going to be produced and partially consumed, with the surplus hydrogen sold as fuel and the surplus electricity sent back onto the energy network, the ability to aggregate large numbers of producer/users into associations is critical to energy empowerment and to the advancement of the vision of democratic energy.

The aggregation of distributed generation has much in common with the aggregation of labor in the early union movement at the beginning of the 20th century. Industrial workers, alone, were too weak to negotiate the terms of their labor contracts with management. Only by organizing collectively as a bloc within factories, offices, and whole industries could labor amass enough power to bargain with management. The ability to withhold labor collectively by using "the strike" gave workers a powerful tool in their campaign to shorten workweeks, improve the conditions of work, and increase both pay and benefits.

By organizing collectively into distributed-generation associations (DGAs), the operators of distributed generation can better dictate the terms with commercial suppliers of fuel cells for lease, purchase, or other use arrangements. The ability of DGAs to aggregate individual fuel-cell operations into vast extended power plants with large power capacity also gives them a leg up with both commercial and noncommercial bundlers of energy, who will help direct and coordinate the flow of hydrogen and electricity to potential consumers on the energy web.

It is unlikely that DGAs will ever go into the business of developing, manufacturing, and marketing fuel cells and the accompanying equipment that goes with distributed generation—although they might enter into joint ventures or purchase stock and secure directorship positions on the boards of commercial enterprises. It is more likely that in the early stages of building local, regional, and global HEWs, DGAs will have to partner with existing power companies, for two reasons: Power companies and transmission companies control existing power grid infrastructures, and they have the expertise to act as "virtual power plants" to coordinate the flow of energy and energy services on the networks.

Power companies are going to have to come to grips with the reality that millions of local operators, generating electricity from fuel cells on-site, can produce more power more cheaply than can today's giant power plants. When the end users also become the producers of their energy, the only remaining role for existing power plants is to become "virtual power plants" that can manufacture and market fuel cells, bundle energy services, and coordinate the flow of energy over the existing power grids.

Putting Theory into Practice

There are a number of existing organizational models that could help establish DGAs. In the United States, community development corporations, credit unions, and public utilities come to mind. In the U.S. and other countries, cooperatives are perhaps the strongest existing model for establishing DGAs. In the developing world, village cooperatives—working with microcredit banks and leveraged by government grants, loans, and development aid—could set up their own DGAs. All of these models are either nonprofit organizations or public entities responsible to their members or, in the case of public utilities, the citizenry. By using not-for-profit organizational models to establish DGAs, the new hydrogen economy would begin with a bold commitment, at the outset of the new energy regime, to the semi-public nature of hydrogen power and distributed-generation networks.

Using existing not-for-profit models of organization to create DGAs and a new energy infrastructure for the 21st and 22nd centuries also has a second effect. DGAs create an appropriate decentralized, bottom-up institutional approach to utilizing the new energy regime that can serve as a foundation for the many other economic, social, and cultural institutions that will accompany the hydrogen economy.

Community development corporations (CDCs) have been in existence for thirty years in the United States and are one of many possible organizational models that could be used to establish DGAs. They are nonprofit organizations, generally established in poor urban neighborhoods and communities (although rural CDCs also exist) whose mission is to stimulate economic development and secure greater control and power for local residents. The late senator Robert Kennedy, who was involved in establishing the first major CDC in the Bedford-Stuyvesant neighborhood of Brooklyn, New York, in 1967, envisioned CDCs as vehicles to combine "the best of community action with the best of the private enterprise system."[9] There are currently between 3,000 and 4,000 CDCs operating in the United States.[10] They are financed, in part, by federal, state, and local government grants and by foundations and commercial institutions. A CDC is governed by a board of directors made up primarily of local residents of the community served by the organization.

In the early years, CDCs concentrated on building affordable housing—creating more than a half million housing units.[11] Today, CDCs have extended their activities to include a wide range of commercial ventures. In Newark, New Jersey, for example, the New Community Corporation owns a two-thirds stake in a Pathmark supermarket and a shopping center and operates day-care centers, a nursing home, a medical day-care center for seniors, a restaurant, a newspaper, and a credit union. Its real estate holdings alone are valued at $500 million. The Newark CDC generates $200 million in revenue annually and is the largest private employer of local residents in the city.[12]

Virtually every one of the 133 largest cities in the United States has at least one active CDC. With their deep roots in residential housing and commercial real estate in some of the poorest neighborhoods in the country, CDCs are ideally positioned to mobilize homeowners and local businesses to lease or purchase distributed-generation systems and create DGAs.

Financing the lease or purchase of fuel cells for homes and businesses could be leveraged, in part, by community development credit unions (CDCUs). There are currently about 1,500 federally insured CDCUs operating in the United States.[13] CDCUs are nonprofit organizations whose depositors are also their members. These community based institutions handle traditional banking services, including holding savings deposits and checking accounts as well as making personal loans for mortgages, home renovations, and the purchase of automobiles. The mission of CDCUs is to provide affordable credit to working and low-income Americans, maintain financial services in poor neighborhoods, and recirculate members' deposits back into the local community. Making loans for the purchase of fuel cells would be in line with CDCUs' stated purpose of advancing the social capital of their respective communities.

Publicly owned not-for-profit utilities (POUs) could also take the lead in establishing distributed-generation associations. POUs were first established more than 100 years ago in the United States to produce electricity, primarily in rural regions of the country that had been passed over by investor-owned utilities that didn't think there was enough of a market to justify the investment. Today, one out of every seven Americans—40 million people—get their electricity from one of the 2,000 public utilities that operate in both rural and urban communities.[14] POUs comprise 12 percent of the country's total installed generating capacity.[15] While three-quarters of the POUs serve small communities of 10,000 or fewer people, some of the largest cities in the country, including Los Angeles, San Antonio, Sacramento, Nashville, Jacksonville, and Memphis, get their electricity from municipally owned utilities. POUs, because they are public not-for-profit institutions, have traditionally provided quality service at lower rates than have commercial utilities. Residential customers pay, on the average, 30 percent more if they receive their power from commercial power plants rather than from POUs. Commercial and residential ratepayers together pay an average of 9 percent more if the electricity is purchased from investor-owned utilities. And only

20 percent of the public power advantage is due to tax-exempt financing. The rest comes from better management and greater efficiency.[16]

Publicly owned utilities, once in decline, are experiencing a mini-renaissance in recent years as commercial and residential customers tire of power outages, blackouts, and price hikes. In the past two decades, forty-five new municipal systems were formed, twenty-three of which were in communities formerly served by investor-owned utilities.[17]

A publicly operated utility, because it is already a not-for-profit institution serving entire communities, could more quickly mobilize its existing customer base into a distributed-generation network, first as an auxiliary or backup to the existing power system, and eventually phased in as an alternative power source. POUs could purchase and install fuel cells in customers' homes and places of business "free" and work out an arrangement whereby surplus power generated by local operators during off-peak load periods could be sent back to the grid, giving them a potential credit on their energy bills. POUs could then sell the surplus energy to buyers in other locales to help pay off the cost of purchasing and installing the fuel cells at the customers' sites and to maintain the energy grid.

There is one more uniquely American institution that, even though it is not a nonprofit institution, could play a major role in establishing distributed generation and laying the groundwork for a national energy web. Today, more than 30 million Americans—12 percent of the population—reside in some 150,000 common interest developments (CIDs).[18] Residents of CIDs own their own units and share ownership of "common areas," including lawns, parks, roads, parking lots, tennis courts, swimming pools, and recreation centers. Every homeowner belongs to a membership association and is required to pay monthly or annual dues for the management and maintenance of the community.

New CIDs are being established at a rate of 4,000 to 5,000 per year.[19] If the current rate of growth continues—and every indication is that it will increase even more rapidly over the next two decades—CIDs could gradually rival existing municipal governments, notes Robert H. Nelson, an economist at the U.S. Department of the Interior.[20]

CID developers could contract with local power companies, before construction, to install a fuel cell in each home or unit free of charge and attach it to the main grid in return for an agreement that the surplus electricity generated would be sold back to the main grid at a discounted rate or provided free to defer the costs incurred by the power company to provide the fuel-cell system and manage the grid. Alternatively, a third-party virtual power plant could finance the cost of purchase and installation of the fuel cells in return for a long-term contract with the membership association that allowed the virtual plant to manage the flow of surplus energy over the main power grid. Already-existing CIDs could decide to phase in backup generators, financing them with annual dues, as a way to ensure the flow of electricity if their electrical service was disrupted. The up-front cost of financing the generators and accompanying equipment could pay for itself and eventually could even generate substantial revenue for the CIDs by selling surplus energy back to the network during peak-load demand.

Globally, cooperatives are perhaps the best organizational vehicle for establishing DGAs. The cooperative movement dates back to the 19th century, when twenty-eight tradespeople in Rochdale, England—angry with the high prices charged by local merchants—banded together to buy food items in bulk at lower prices and then resold them back to their members at lower-than-retail prices. The group, the Rochdale Equitable Pioneers Society, drew up a mission statement, later known as the Rochdale Principles, that continues to be used, in moderated form, by co-ops around the world.[21] According to the International Cooperative Alliance (ICA), headquartered in Geneva, Switzerland, co-op principles include inclusive membership, democratic participation, equitable distribution of resources, autonomy, education, cooperation among cooperatives, and concern for community. The ICA claims a membership of 230 organizations, made up of 750,000 trade and consumer groups, operating in 100 countries, with a total membership of 730 million people around the world.[22]

The corporate sector in the U.S. likes to think that the American economy is run almost exclusively by private enterprise. That is not altogether so. There are more than 48,000 cooperatives in the U.S., and they account for more than $125 billion in annual economic activity.[23] According to the National Cooperative Business Association (NCBA), more than 75 million Americans bank at cooperative credit unions, 50 million Americans hold insurance policies owned by or affiliated with cooperatives, and 34 million Americans get their electricity, at cost, from rural electric cooperatives. Thirty percent of all farm products are marketed through cooperatives. Cooperatives also provide health insurance for 1.4 million people, and more than 1 million households are parts of cooperatives.[24] National brands like Ace Hardware, Land O' Lakes, ShopRight, Ocean Spray, Sunkist, and REI are all cooperatives.[25]

Cooperatives are popular because they allow an aggregation of individual producers or consumers to bargain more effectively with suppliers. Producers are especially attracted to cooperatives because they can pool their finances and risks and establish more effective shipping and marketing channels. Cooperatives also afford their members a sense of direct participation and control over the enterprise. "One member, one vote" is still the general operating principle in the ownership of co-ops, as opposed to that in investor-owned enterprises, in which the number of votes is fixed by the amount of the stock holding.

In many countries, nonprofit cooperatives are among the biggest and most powerful commercial enterprises. Mondragon in Spain is made up of 160 employee-owned enterprises operating in several countries with annual combined sales of more than $5 billion and assets of more than $10 billion (1999 figures).[26] Italy boasts 250,000 nonprofit worker cooperatives made of small- and medium-sized crafts and industrial companies.[27] The Cooperative Group of the U.K. includes more than 1,100 food stores, a cooperative bank, an insurance cooperative, a farm-care cooperative, the largest funeral service company in Britain, a cooperative retail group that manufactures its own brand names and manages more than 1,900 retail stores, the nation's fourth-largest travel agency, and 57 eye-care practices.[28] Japan's consumer cooperative has 19 million members—20 percent of Japanese households.[29]

With 730 million members in 100 countries, cooperatives could help lead the way into a hydrogen era by establishing distributed-generation associations in thousands of communities. Because cooperatives are already largely geographically based, made up of an aggregate of individual producers or consumers, bound together as membership organizations, and nonprofit in practice, they fit well with the requisites of a distributed-generation energy web. They provide a bottom-up organizational structure to go along with a decentralized bottom-up energy infrastructure, and they provide a democratic form of governance to help facilitate the democratization of energy in the new hydrogen era.

In the U.S., Touchstone Energy, a national alliance of 550 consumer-owned electric cooperatives that provides electric power to sixteen million customers in thirty-nine states, has established, for the first time, a nationwide network to connect all of the electric cooperatives. The new network buys advertising on television and in the print media, provides a bill-consolidation and energy-management program for its member co-ops as well as coast-to-coast call centers to enable cooperatives to share resources across time zones and provide customer coverage around the clock. This kind of nationwide network could be used to help generate popular public support for distributed generation and the wiring of a nationwide hydrogen energy web.[30]

The recently organized Energy Cooperative Development Program, run by the California Energy Commission, is one of several new prototype organizations that could help establish DGAs around the country. The program's goal is to help consumers organize into energy cooperatives and to create a regional and national cooperative energy network. The initial target audience includes small businesses, small agricultural firms, moderate- and low-income residential customers, seasonal growers, renters, and the elderly on fixed incomes, as well as Native Americans and other minorities whose primary language is not English.[31]

Two electric-user cooperatives, "utilities without wires," have already been formed, one in California and the other in New York City. The California Electric Users Cooperative is a federation of eighteen agricultural cooperatives that have entered into a full-service electric-supply contract for electricity with New West Energy of Phoenix, Arizona.[32] The arrangement is a first in the energy business. The 1st

Rochdale Cooperative Group Inc. in New York City is the first electric-user cooperative to serve a major U.S. city.[33] In the mid-1990s, concerned that electricity rates in New York were among the highest in the nation and a major cost factor in the budgets for housing cooperatives, and aware that the restructuring of the electricity industry was opening up opportunities for independent ventures, 1st Rochdale sought out the expertise of the National Rural Electric Cooperative Association (NRECA) and created an alliance of mutual interest. Today, 1st Rochdale is buying energy in bulk and saving energy costs for 50,000 New York City apartments. If many of the 500,000 New York families living in cooperative apartments were to join 1st Rochdale, it would become a powerful player in the national energy game and a model for similar programs elsewhere.[34] 1st Rochdale has already taken its first step into distributed generation and the creation of an energy web. Its Green Apple initiative promotes both the use of sustainable energy and distributed generation.

In Chicago, the newly established Community Energy Cooperative has joined with the Electric Power Research Institute (EPRI) in a pilot project to provide distributed generation to several Chicago-area families. Small fuel-cell power plants are being installed in the homes to produce the residents' power needs. According to Dan Restler, EPRI's area manager for distributed resources, the pioneering effort "is a step toward decentralized mini-grids where communities might be partially served by residential-size or larger distributed resources."[35]

The Community Energy Cooperative is the brainchild of the Center for Neighborhood Technology, a Chicago-based nonprofit organization committed to building sustainable communities. The cooperative was launched with the help of ComEd, a Chicago-based utility and one of the biggest power companies in the United States. ComEd provided a three-year startup grant to help get the cooperative off the ground. The cooperative is working with its members to reduce energy use during peak-load demand. ComEd, in turn, pays the cooperative for the energy savings. Some of the payment is used by the cooperative to offer discounts to its members on energy-efficient appliances and technologies—which, in the near future, will also include the purchase, lease, and installation of fuel-cell power plants in members' homes and businesses. The rest of the ComEd payment is used by the cooperative to support youth programs, playgrounds, senior centers, and other community programs. The cooperative expects to invest more than $1 million in energy-saving revenue in the local communities that it serves by the end of 2003.[36]

The joint effort by the Community Energy Cooperative, ComEd, and the EPRI is a model for the kind of partnerships that are likely to form in the future between cooperatives and other nonprofit community-based organizations and the nation's power companies. Already, plans for establishing similar energy cooperatives along the lines pursued by 1st Rochdale and Chicago's Community Energy Cooperative are being actively pursued in major U.S. cities, including Washington, D.C., and Philadelphia.[37]

Many of the structural elements necessary to make a HEW viable are already in place in the U.S. and other countries. If the existing power structure becomes more vulnerable to disruption, power outages, and blackouts, and if the cost of buying electricity continues to rise, expect new energy players, especially not-for-profit cooperatives, municipal utilities, neighborhood associations, community-development corporations, community credit unions, common-interest developments, and others to leap into the breach, creating joint collaborative ventures, both with one another and commercial providers, to establish distributed-generation associations and begin to erect hydrogen energy webs.

Empowering the Poor

Sixty-five percent of the human population have never made a single telephone call, and one-third of the human race has no access to electricity or any other form of commercial energy.[38] The disparity between the connected and the unconnected is deep and threatens to become even more pronounced over the next half century with world population expected to rise from the current 6.2 billion to 9 billion people.[39] Most of the population increase is going to take place in the developing world, where the poverty is concentrated.

Although many well-off American consumers continue to labor under the illusion that the divide between rich and poor is narrowing in the world with steady advances in science, technology, and commercial innovations, the opposite is true. More than 100 countries, with a combined population of 1.6 billion people, are experiencing economic decline. And eighty-nine countries are worse off now in per-capita income than they were a decade ago. In Africa, the average household actually consumes 20 percent less today than it did twenty-five years ago.[40] Meanwhile, the International Labor Organization estimates that one-third of the 3 billion workers in the world are currently either unemployed or underemployed.[41]

Abject poverty is the ever-present reality throughout much of the world. Currently, 600 million people are homeless or living in unsafe housing, and the World Bank estimates that, by 2010, 1.4 billion people will live without clean water and sanitation.[42] While the top 20 percent of high-income earners in the world now account for 86 percent of the private consumption, the poorest 20 percent consume less than 1.3 percent of global economic output.[43] Even more incredible, according to a study conducted by the United Nations Development Program (UNDP), the 358 richest people in the world have combined assets that exceed the total annual income of nearly half the people on Earth.[44]

Lack of access to energy, and especially electricity, is a key factor in perpetuating poverty around the world. Conversely, access to energy means more economic opportunity. In South Africa, for example, for every 100 households electrified, ten to twenty new businesses are created.[45] Electricity frees human labor from day-to-day survival tasks. Simply finding enough firewood or dung to warm a house or cook meals in resource-poor countries can take hours out of each day. Electricity provides power to run farm equipment, operate small factories and craft shops, and light homes, schools, and businesses.

As mentioned earlier, the amount of energy consumed per capita is a litmus test, throughout history, of human beings' ability to advance beyond a survival existence. Today, the per-capita use of energy throughout the developing world is a mere one-fifteenth of the consumption enjoyed in the United States. The global average per-capita energy use for all countries is only one-fifth the level in the United States.[46]

In his study of social categories and electricity use, Chauncey Starr, president emeritus of the Electric Power Research Institute, observes that below a critical threshold of annual per-capita income and energy consumption, human beings are forced to spend each day concentrating almost exclusively on their "survival"—that is, just finding temporary shelter, enough food to eat, and enough water to drink. Once people secure minimum employment and some access to electricity, they advance to a "basic quality of life," including literacy, better sanitation, physical security, and a longer life expectancy. As per-capita income and electricity consumption increase, people can move up to an "amenities" lifestyle that includes education, recreation, and intergenerational investment. At the top rung of social categories, what Starr calls "international collaboration," people enjoy unlimited access to electricity and can create communities of interest that are truly global in scale.[47] The problem is that in the next half century more than 90 percent of the world's population will be born into the "survival" or "basic quality of life" social categories.[48]

While one-third of the world's population has no access to electricity, the situation isn't much better for those who do. Today, half the human race exists on less than a few hundred kilowatt-hours of electricity per person per year—hardly enough to make much difference in their quality of life.[49] To achieve universal global electrification by 2050—a goal set by international development agencies—would require bringing electricity to 100 million additional people per year, more than two and one-half times the number of new electricity users brought in currently each year. Providing these additional 100 million new users with an average per-capita consumption of electricity equivalent to what U.S. consumers enjoyed in 1950 would require the creation of 10 million megawatts of new electricity capacity globally by 2050—that is, four times today's consumption.[50] The Electric Power Research Institute (EPRI) estimates that to reach this goal, a new 1,000-megawatt power plant would have to be brought on line every forty-eight hours for the next fifty years.[51] As if this weren't challenging enough, the EPRI adds that 50 percent of the new capacity would need to be carbon-free to comply with global environmental requirements. Getting the job done would necessitate a capital commitment of between $100 and $150 billion per year.[52] The International Energy Agency calculates that new power generation in developing countries alone will need to be on the order of $1.7 trillion between 1995 and 2020.[53]

The question is, how realistic is universal electrification when global energy per capita has already peaked and is moving down the backside of the bell curve, and when global oil and natural-gas production are expected to peak some time before 2020? Even the EPRI says that the effort "will not be easy."[54] That is because the switchover to natural gas to produce electricity, in the U.S. and in other countries, cannot carry most of the load much beyond 2030. As mentioned earlier, the EPRI estimates that as natural gas is relied on more and becomes increasingly scarce, prices are likely to rise dramatically—by as much as 50 percent.[55] After 2030, says the EPRI, we will have to look to other strategies. The only other viable strategy left is to use renewable energy resources to create a hydrogen energy regime.

Even here, the task is daunting. To create a sustainable future that could provide sufficient energy for a decent quality of life for every human being on Earth, while reducing harmful global warming emissions, would require a growth in productivity of about 2 percent a year. We can see how difficult this will be to accomplish when we stop to consider that, while the U.S. was able to achieve a cumulative 2 percent advance in productivity—double the world average—over the past century, it was able to do so largely because it had access to a vast supply of cheap crude oil.[56]

Making the shift to a hydrogen energy regime—using renewable resources and technologies to produce the hydrogen—and creating distributed-generation energy webs that can connect communities all over the world is the only way to lift billions of people out of poverty. Narrowing the gap between the haves and have-nots means first narrowing the gap between the connected and the unconnected.

As the price of fuel cells and accompanying appliances continues to plummet with new innovations and economies of scale, these products will become far more broadly available, as was the case with transistor radios, computers, and cellular phones. The goal ought to be to provide stationary fuel cells for every neighborhood and village in the developing world. Villages can install renewable energy technologies—photovoltaic, wind, biomass, etc.—to produce their own electricity and then use the electricity to separate hydrogen from water and store it for subsequent use in fuel cells. In rural areas, where commercial power lines have not yet been extended because it is too expensive, stand-alone fuel cells can provide energy quickly and cheaply. After enough fuel cells have been leased or purchased and installed, mini-energy grids can connect urban neighborhoods, as well as rural villages, into expanding energy networks. The HEW can be built organically and spread as the distributed generation becomes more widely used. The larger hydrogen fuel cells have the additional advantage of producing pure drinking water as a by-product, a not-insignificant consideration in village communities around the world where access to clean water is often a critical concern.

No longer will Third World nations have to be dependent on the flow of crude oil. Recall that the 1970s OPEC price spikes had a far more adverse effect on the poor countries of the world than on the industrialized countries. In the aftermath of the OPEC oil embargo of 1973, the price of oil quadrupled from $3 to $13 per barrel on world markets, wreaking havoc on developing nations. Third World countries were forced to seek loan assistance from Western banks and international lending institutions like the World Bank and the International Monetary Fund (IMF) to pay for the increased cost of oil imports, without which their economic-development projects would have stalled. Commercial bank loans to the Third World increased by 550 percent between 1973 and 1980.[57]

The second spike in OPEC oil prices in 1979 triggered a global recession and a collapse of commodity prices, further weakening debt-ridden developing countries. The escalating costs of oil imports, combined with falling prices for their commodities in world markets, forced developing countries to borrow still more funds—most of which were used merely to buy oil and pay off the interests on past debts. By 1985, Third World debt exceeded $1 trillion.[58] Since most of the funds they were borrowing were being used to buy oil and repay past loans, little money was left over for real economic development. The result was that Third World nations lost economic momentum and slipped further into poverty. Commercial and institutional credit began to dry up, sending Third World countries into a downward economic spiral. By 1988, many developing countries were experiencing a net loss of money.[59]

Conditions have only worsened in recent years, as developing countries have become even more dependent on oil imports to provide power for their expanding manufacturing sector, light and heat for an increasingly urban population, and fuel for their growing transportation needs. In the 1970s, developing countries accounted for only 26 percent of total world oil demand. Today, their share tops 40 percent and is continuing to climb.[60] In 2000, the rise in oil prices added $6 billion to the import bill in India alone. Brazil's oil import bill was 150 percent higher in 2000 than in 1999. China experienced a 250 percent increase in its oil import bill in 2000.[61] In many countries, the increased costs of securing oil has canceled the benefits of international-development assistance. According to the EIA, the additional cost of imported oil is two and one-half times the amount given as foreign-aid assistance in countries like China and Thailand.[62]

The Secretary General of the United Nations, Kofi A. Annan, recently warned that, for most developing countries, "debt-servicing costs are likely to increase if higher oil prices lead to higher international interest rates" in the years ahead.[63] Third World debt has already reached crisis proportions, even though global oil production has yet to peak. By the end of 1999, forty-seven countries—with a combined population of 1.1 billion people—owed more than $422 billion to foreigners. The average per-capita debt in these countries is $380, an amount roughly equal to the average per-capita gross national product.[64] Even more important, for the poorest countries—those with GDPs that are less than $885 per year per capita—eighty-three cents of every dollar received in new loans is being used merely to pay back old loans, leaving little left to spur development or raise living standards.[65] The human dimension of the debt crisis is staggering. Many of the poorest countries in the world now spend more of their domestic revenue on servicing their foreign debts than on basic health, education, and social services.[66]

It is not surprising, then, that anti-globalization protesters at world-development forums have made the issue of cancellation of Third World debt a rallying cry. While creditor institutions have agreed to cancel a small portion of the mounting debt, critics argue that this will not be nearly sufficient to reverse the worsening economic plight of most Third World nations.[67] Only by freeing themselves from dependence on foreign oil and gas imports can Third World countries get out from under and improve the economic conditions of their populations.

Distributed-generation associations (DGAs) need to be established throughout the developing world. Civil society organizations (CSOs), cooperatives (where they exist), microcredit lending institutions, and local governments ought to view distributed-generation energy webs as the core strategy for building sustainable, self-sufficient communities. "Empowerment," in this context, takes on a double meaning. Without access to energy, and particularly to electricity, people will remain powerless to control their own personal destinies. Breaking the cycle of dependency and despair—becoming truly "empowered"—starts with access to and control over energy.

National governments and world lending institutions need to be lobbied or pressured to help provide both financial and logistical support for the creation of a hydrogen-energy infrastructure. Equally important, new laws will need to be enacted to make it easier to adopt distributed generation. Public and private companies will have to be required to guarantee distributed-generation operators access to the main power grid and the right to sell energy back or trade it for other services.

The fossil-fuel era brought with it a highly centralized energy infrastructure and an accompanying economic infrastructure, both of which favored the few over the many. Over the course of the past century, the deepening divide between the haves and have-nots, and now the connected and unconnected, has been attributable in no small part to the nature of our fossil-fuel energy regime.

Now, on the cusp of the hydrogen era, it is possible to imagine a decentralized energy infrastructure, the kind that could support a democratization of energy, enabling individuals, communities, and countries to claim their independence while accepting responsibility for their interdependence as well.

In the early 1990s, at the dawn of the Internet era, the demand for "universal access" to information and to communications became the rallying cry for a generation of activists, consumers, citizens, and public leaders. Today, as we begin our journey into the hydrogen age, the demand for universal access to energy ought to inspire a new generation of activists to help lay the groundwork for establishing sustainable communities. If that were to happen, we could begin to reglobalize power along wholly new lines. This time, power would flow laterally from home to home, neighborhood to neighborhood, and community to community, creating a vast decentralized energy infrastructure that would promote both the values of self-sufficiency and interdependence.

Were all individuals and communities in the world to become the producers of their own energy, the result would be a dramatic shift in the configuration of power: no longer from the top down but from the bottom up. Local peoples would be less subject to the will of far-off centers of power. Communities would be able to produce many of their own goods and services and consume the fruits of their own labor. But, because they would also be connected via the worldwide communications and energy webs, they would be able to share their unique commercial skills, products, and services with other communities around the planet. This kind of economic self-sufficiency becomes the starting point for global commercial interdependence and is a far different economic reality than that in colonial regimes of the past, in which local peoples were made subservient to and dependent on powerful forces from the outside.

Economically sustainable local communities make possible more than just material well-being. Empowering local communities also helps preserve the rich cultural diversity of the human family. Economic self-sufficiency provides the material security that a people needs to maintain a sense of social cohesion and to preserve its cultural largesse. At the same time, embeddedness in larger global communications and energy networks frees people from the xenophobia that traditionally accompanied a more isolated geographic existence. In the new global context, local culture becomes less of a possession to defend and more of a gift to share with the world. Cultural exchange once again reasserts itself and becomes as powerful an expression of human interaction as commercial exchange. Social capital flourishes alongside market capital, and political power emanates from deep inside the culture rather than from the commercial or governmental spheres.

In the new scheme of things, globalization starts with the full empowerment of each human being and radiates out to the family, the community, the country, the networks of commercial interest, and the biosphere itself. By redistributing power broadly to everyone, it is possible to establish the conditions for a truly equitable sharing of the Earth's bounty. This is the essence of the politics of reglobalization from the bottom up.

Rethinking Security

The hydrogen economy is going to bring with it a new way of thinking about the sociology of our existence, just as the fossil-fuel era did. The great transformations in energy regimes throughout history have always forced a rethinking of the most basic categories of human existence. Hunter-gatherer cultures, although each unique, share a commonality of spirit, as do agricultural and industrial societies. The way human beings gather, transform, and use the dominant forms of energy at their disposal—whether it be wild animals and plants; domesticated food crops; human slaves; or coal, oil, and natural gas harnessed to machines—becomes imprinted in the various ways they define their ideas about personal and collective security.

For example, consider the notion of security held by Christians in the late medieval era with that of the bourgeoisie of 20th-century industrial capitalism. The two very different ideas about security tell us a great deal about the very different natures of the energy regimes that people have depended on for their existence.

As we mentioned earlier in the chapter, medieval Europe was loosely organized under the aegis of the Catholic Church, in consort with local warlords—kings, princes, and manor lords. Society was conceived of as an intimate microcosm of God's grand creation, which descended from heaven in the form of a great ladder or chain of being with God on top; and below him his emissaries on Earth, the pope and priesthood; and then lesser personages, including kings, lords, knights, farmers, tenants, and serfs; and finally our fellow creatures all the way down to the lowest regions and to every "creeping thing that creepeth upon the Earth." Every rung of the divine ladder, explained St. Thomas Aquinas, was occupied by one of God's creatures, and all the rungs were full, leaving no room for innovation, surprises, or changes in God's master plan. The Church's world was a tightly knit structure with a careful gradation of ranks and a detailed catechism of instructions governing mutual obligations. Security depended on human beings executing the tasks assigned to them by Providence and accepting their roles and responsibilities in the natural hierarchy. By faithfully fulfilling their obligations and duties in this world, medieval men and women could be assured a modicum of security on Earth as well as everlasting security in the next world.

The land, in the medieval scheme of things, was particularly important in defining security, as it was the place where one served as a steward of God's creation. In an era when agriculture was the dominant energy regime, security naturally flowed from the land. The sense of place was all the more poignant because in the feudal order people, for the most part, belonged to the land, rather than the land belonging to the people. Security, then, was a vertical affair that began in this world with a sense of attachment to the ancestral home, to which one was born, and ascended up to the next world, where one was rewarded with everlasting life.

The coal economy fundamentally changed the security equation. Now steam-driven machines allowed society to augment human slaves and animals with mechanical substitutes. The new machine-slaves made "man" feel less dependent on God and the forces of nature for his well-being. Autonomy gradually replaced divine deliverance, at least for those who controlled and benefited from what historian Arnold Toynbee later called "The Industrial Revolution."

The coal economy also greatly quickened the pace, speed, flow, diversity, and intensity of human life. Whereas, during the Middle Ages, few people wandered much beyond a day's walk from their homes, by the late 19th century, millions of people were regularly crossing oceans and whole continents by steamships and trains in just a few days. The railroad and telegraph shortened distances, compressed time, and added a new dimension to human life—"mobility." Millions of farmers, uprooted from their ancestral lands and feudal bondage, migrated to sprawling, bustling cities where they became "free labor," working in the factories alongside steam-driven machinery. The new urban age was increasingly characterized by a sense of restlessness and perpetual change. Innovation became the hallmark of the industrial economy. To be modern was to be avant-garde, to be continually experimenting with new ideas, fashions, and lifestyles. A younger generation castigated its elders for being too "old-fashioned."

The growing sense of self-reliance and the quickened pace of life that so marked the new era evolved into a wholly new idea about security. The medieval view of security that was bound up with belonging to the land in this world and being saved in the next world was gradually replaced with a new sense of security based increasingly on the notion of individual autonomy and mobility. These two values were to become the dominant virtues of an emerging bourgeois class in every industrial country.

Autonomy became synonymous with freedom. To be autonomous was to be in control of one's destiny and not dependent on others. The key to gaining one's autonomy was to be propertied. The new role of government, in turn, was to defend and expand markets and thus assure everyone's right to amass property so that they could gain their autonomy and be truly free.

Mobility, in the new era, meant far more than safe passage. To be mobile was always to have new choices and options available. Just as autonomy became synonymous with freedom, mobility became synonymous with opportunity. The fossil-fuel era freed human beings from the slower seasonal rhythms of an agricultural period and thus also from dependency on nature's constraints and divine intervention. Each new convert to modernity was endowed with a Promethean spirit. Armed with autonomy and mobility, every human being could become a minor god, securing a small, personal piece of paradise in this material world.

By the 20th century, the middle class, whose sense of security was so bound up in both autonomy and mobility, found the ultimate object of their affection in the invention of the "auto-mobile." Here was a machine, fueled by gasoline, that in one stroke provided a complete sense of autonomy and mobility for its driver and passengers. It is no wonder that it so quickly became the prime metaphor for the age and the centerpiece of the industrial economy. Henry Ford once mused that his car should be regarded as a "living room on wheels." What youngster has not experienced the rush of freedom and opportunity that comes the first time he or she takes the driver's seat and puts hands on the wheel and a foot to the accelerator? In the car, on the road, one feels not only both autonomous and mobile but also as secure as it is possible to be in the modern era, and all this was made possible by the harnessing of fossil fuels.

The modern idea about security provided a framework for countries as well as for individuals. The new "science" of geopolitics, first outlined in 1882 by Friedrich Ratzel, a German professor of geography, became a mirror image of the new bourgeois sensibility.[68] Borrowing heavily from Darwin's theory of the origin and development of species, with its emphasis on biological competition for scarce space and resources, Ratzel argued that, in seeking to enlarge its commercial opportunities and military rule, the nation-state was merely fulfilling its biological destiny. Nations seek to secure their autonomy by continually expanding their reach, and the key to their success is the degree of their mobility. Ratzel believed that mobility could best be realized by maintaining a commanding presence on the world's oceans. "Germany must be strong on the seas," wrote Ratzel, "to fulfill her mission in the world."[69]

Sir Halford Mackinder, a British geographer, agreed with Ratzel that control over the world's oceans, which make up nine-twelfths of the surface of the Earth, would give any nation the mobility it needed to fend off aggressors while allowing it to advance its own expansionist plans. Later, after World War II, geopolitical theorists extended their thinking to include mobility in the air as well.[70]

Architects of the new geopolitics were quick to emphasize the strategic importance of coal, oil, iron, and other minerals, without which modern warfare could not be fought and an industrial way of life could not be advanced. Nicholas Spykman, one of the leading geopolitical figures of the early 20th century, wrote that control over natural resources was a critical factor in formulating foreign policy.[71] As we learned earlier, political and military leaders on both sides of the two world wars came to understand exactly how important securing oil and other vital resources was to achieving autonomy, mobility, and victory.

Modern notions of security and geopolitics made sense in a world in which time was still linear and space was still expansive. The temporal and spatial playing field was still big enough to warrant putting a premium on autonomy and mobility.

But, in a world in which technology allows people to interact with one another and conduct business and social life at the speed of light, 24/7, duration narrows to near simultaneity, and distances nearly disappear. The Earth becomes less of a field of encounter and more of an indivisible entity. The first realization of the new reality came in 1945 when the U.S. military dropped atomic bombs on the Japanese cities of Hiroshima and Nagasaki. Suddenly, humanity came to understand that it wielded sufficient power to destroy itself. We began to talk, for the first time, about living in a single world in which everyone's security was either mutually secured or lost. The very idea of somehow being able to secure individual or national autonomy in a nuclear world seemed quaint and utterly naive.

The first photos of the Earth taken from outer space in the 1960s also helped change our ideas about security. We saw ourselves, in a visceral way, as collectively inhabiting a tiny sphere revolving around a small star somewhere in the giant universe. Whereas autonomy assumes the one against the many and mobility assumes vast space in which to maneuver, the new view of Earth from a distance suggested that there was only a single Earth community confined to a very small space in the cosmic theater.

Perhaps the most compelling new reality that has forced us to rethink our conventional notions of security is the rise in the Earth's temperatures caused by the burning of fossil fuels during the Industrial Age. We spent the better part of the fossil-fuel era attempting to enclose the Earth's ecosystems into commercial arenas of engagement so that we could advance our individual and collective sense of autonomy and mobility. Now the Earth is enclosing us in our own spent energy, undermining both the autonomy and the mobility that we so desperately sought during the whole of the modern age.

We now find ourselves caught up in a new scientific, technological, and commercial revolution that can, in theory, connect the whole of humanity and our fellow creatures into a single, indivisible web of existence. Yet our ideas about security are still wedded to an age based on the narrow Darwinian assumptions that each organism is an island seeking only to optimize its mobility and secure its autonomy.

There are signs that the first generation to grow up with videogames, computers, palm pilots, and cell phones, and who spend so much of their time connecting with one another and sharing information on the Internet and the World Wide Web, may be developing a different kind of consciousness and, with it, a new sense of security. For the dot-com generation, the very idea of autonomy, which so affected the thinking about security for their parents' generation, is suspect, if not just plain irrelevant to their needs in a wired world. Whereas previous generations defined freedom in terms of autonomy and exclusivity—each person is a self-contained island—the children of the Web have grown up in a very different technological environment, in which autonomy is thought of (if at all) as isolation and death, and in which freedom is more likely to be viewed as the right to be included in multiple relationships. Their identities are far more bound up in the networks to which they affiliate. For them, time is virtually simultaneous, and distances hardly matter. They are increasingly connected to everyone and everything by way of an electronically mediated central nervous system that spans the whole of the Earth and seeks to encompass virtually everything in it. And, with each passing day, they become more deeply embedded in a larger social organism, in which notions of personal autonomy make little sense and the feeling of unlimited mobility is circumscribed by the sheer density and interactivity that bind everyone so tightly together.

The hierarchical commercial networks, with their centralized command-and-control systems that went hand in hand with a fossil-fuel energy regime and an industrial way of life, were designed with the modern notion of security in mind. The corporate enterprises that rule the global economy in the last stages of the hydrocarbon age are a paean to the idea that institutional security is best advanced by pushing for ever greater autonomy and mobility.

Now, however, the worldwide communications web and the prospective hydrogen energy web make possible the reorganization of commercial life. In the new era of decentralized command-and-control mechanisms and potentially democratic communications and energy utilization, in which everyone is increasingly linked to everyone and everything else in multiple networks of engagement, at the speed of light, we can begin to rethink our ideas about security.

From Geopolitics to Biospherepolitics

New scientific understandings of how the Earth functions provide a unifying framework for rethinking the question of security in the hydrogen era. The first scientist to think of the Earth as a "living organism" was a Russian, Vladimir Vernadsky. In 1926, Vernadsky wrote a now-famous book, *Biosfera*, in which he hypothesized—contrary to Darwin's theory, which argued that geochemical processes evolved first and created the atmospheric environment in which living organisms could emerge—that in fact, geochemical and biological processes evolved together in a symbiotic relationship. Vernadsky believed that the cycling of inert chemicals on Earth is influenced by the quality and quantity of living matter, and that living matter, in turn, influences the quality and quantity of inert chemicals being cycled throughout the planet.[72]

The biosphere is the thin envelope that stretches thirty to forty miles from the depths of the ocean to the upper stratosphere and contains all the forms of life that exist on Earth. Within this narrow vertical band, living creatures and the Earth's geochemical processes interact to sustain life.

In the 1970s, an English scientist, James Lovelock, and an American biologist, Lynn Margulis, expanded on the Vernadsky thesis with the publication of the Gaia hypothesis. They argued that the Earth functions like a self-regulating living organism. According to their theory, the biota (the flora and fauna of a given region) and the geochemical composition of the atmosphere work in a symbiotic relationship to maintain the Earth's climate in a relatively steady state that is favorable to life.

Lovelock points to the regulation of oxygen and methane as

prime examples of how the cybernetic process between life and the geochemical cycle works to maintain a homeostatic climatic regime on Earth. He reminds us that oxygen levels on the planet must be confined within the narrowest of ranges. A 1 percent rise in oxygen level on the planet would increase the likelihood of fire by 70 percent. A 4 percent rise would likely engulf the entire planet in flames, resulting in the complete destruction of living matter on the land surface.[73] Oxygen production is maintained by photosynthesis. The green chloroplasts inside plant cells convert the sun's energy into chemical energy for the plant's nurturance and, in the process, convert carbon dioxide and water into oxygen. Animals, in turn, take in the oxygen, sustaining their lives, and emit carbon dioxide into the environment. Much of the carbon dioxide is then cycled back through the plant chain again, and so on.

While scientists have known for quite a while how the oxygen and carbon dioxide cycles interact, they have been in the dark as to how the oxygen levels remain so fixed despite major changes in the sun's output and the kinds and number of living creatures inhabiting the planet. To understand how the oxygen level remains fixed at 21 percent, it is necessary to understand how it reacts with other atmospheric gases. Lovelock uses methane to partially explain the process.[74]

It was less than thirty years ago that scientists realized that methane was a biological by-product produced by bacterial fermentation. Microorganisms living inside ruminant animals, in termites, and in peat bogs produce more than 1,000 million tons of methane per year.[75] Methane migrates to the atmosphere, where it acts as a regulator, both adding and taking away oxygen from the air. As it reaches the stratosphere, methane oxidizes into carbon dioxide and water vapor. The water, in turn, separates into oxygen and hydrogen. The oxygen descends to Earth, while the hydrogen escapes into outer space. Methane, then, can add to the existing levels of oxygen in the upper atmosphere. In the lower atmosphere, the oxidation of methane uses up oxygen—about 1,000 megatons each year. Lovelock points out that, "in the absence of methane production, the oxygen concentration will rise by as much as 1 percent in as little as twenty-four thousand years: a very dangerous change, and on the geological timescale, a far too rapid one."[76]

Lovelock and Margulis believe that when the oxygen in the atmosphere rises above a tolerable level, a warning signal of some kind triggers an increase in methane production by microscopic bacteria. The increased methane migrates into the atmosphere, dampening the oxygen content until a steady state is reached again. The constant interaction and feedback between living creatures and the geochemical content and cycles act as an organic unity, maintaining the Earth's climate and environment and preserving life.

> It is an intriguing thought that without the assistance of these anaerobic micro-flora living in the stinking muds of the seabeds, lakes, and ponds, there might be no writing or reading of books. Without the methane they produce, oxygen would rise inexorably in concentration to a level at which any fire would be a holocaust, and landlife, apart from micro-flora in damp places, would be impossible.[77]

Much of what makes up the biosphere, then, either comes from living creatures or is modified by them. Lovelock reminds us that the oxygen and nitrogen of the air come directly from plants and microorganisms. Chalk and limestone deposits are the shells or bones of once-living marine animals. Coral reefs and many islands are merely the burial grounds of untold animalcules. Lovelock cites these and other examples to drive home the point that life is not simply added to a static, inert world of matter "determined by the dead hand of chemistry and physics." Rather, says Lovelock, "the evolution of the rocks and the air and the evolution of the biota are not to be separated."[78]

The planet, then, is more like a living creature, a self-regulating organic entity that maintains itself in a steady state conducive to the continuance of life. According to the Gaian way of thinking, the adaptation and evolution of individual creatures become part of the larger process: the adaptation and evolution of the planet itself. It is the continuous symbiotic relationship between every living creature and the geochemical processes that ensures the survival of both the planetary organism and the individual species that live within its biospheric envelope.

Many other scientists have since weighed in on the Gaia thesis, moderating, qualifying, and expanding on Lovelock and Margulis's work. For more than two decades, the idea that the Earth functions as a living organism has become a critical avenue of departure for rethinking the relationship between biology, chemistry, and geology.

If, in fact, the Earth does function as a living organism, then human activity that disrupts the biochemistry of that organism can lead to grave consequences, both for human life and the biosphere as a whole. The massive burning of fossil-fuel energy is the first example of human activity, on a global scale, that now threatens a radical shift in the climate of the Earth and the undermining of the biosphere that sustains all living creatures.

Our dawning awareness that the Earth functions as an indivisible living organism requires us to rethink our notions of security. If every human life, the species as a whole, and all our fellow creatures are entwined with the geochemistry of the planet in a rich and complex choreography that sustains life itself, then we are, each and all, dependent on and responsible for the health of the whole organism. Carrying out that responsibility means living out our individual lives, in our neighborhoods and communities, in ways that promote the general well-being of the larger biosphere within which we dwell.

French scientist Rene Dubos captured the new sense of security that emerges from our new scientific understanding of how the Earth functions when he urged his fellow human beings "to think globally and act locally." The Internet and the creation of the World Wide Web give us an electronically mediated "central nervous system" to connect everyone and everything within the biosphere so that we can, for the first time, really "think globally and act locally." At the same time, the hydrogen energy web gives us a new, non-polluting energy regime that decentralizes and democratizes energy so that human populations can live in smaller, more dispersed communities that are less likely to stress the biosphere beyond its limits. The mass depopulation of the countryside and the migration to mega-cities with populations in the millions—which is the defining social feature of the fossil-fuel era—are simply unsustainable from a thermodynamic perspective. Because hydrogen energy is ubiquitous and fuel-cell power plants can be put anywhere and connected in extended energy webs, we can bypass the hierarchical and centralized architecture of the oil age. In the hydrogen economy, industrial and commercial activity can be spread out in a more ecologically sustainable manner, allowing for a more balanced mix in the density of human settlement.

Moreover, the universal availability of hydrogen and the lightness, flexibility, and eventual cheapness of decentralized distributed-generation technologies will inevitably raise the question of whether existing political institutions and boundaries ought to be rethought. The nation-state, after all, is a unique creation of the fossil-fuel era. As we learned in chapter 4, the introduction of the coal-powered steam locomotive and the telegraph connected commerce and social life over much wider terrains, giving rise to the nation-state, a kind of "polity of scale" that could control vast geographic regions. Unfortunately, the redrawing of political boundaries into nation-states often bore little relation to ecosystem dynamics, making it difficult for human populations to live in a sustainable way.

In the hydrogen economy, with its decentralized and democratized energy web, it is possible to establish human settlements by bioregions, ecoregions, and georegions that reflect the settlement patterns of the many bio-chemical communities of the Earth itself. Embedding human communities into biocommunities creates a deep new sense of security that is indivisible from the Earth's own health and well-being.

By creating an economic and social architecture that is a microcosm of the Earth's own diverse physiology, we bring our species into a world full of new possibilities that are both life-affirming and regenerative in nature. We finally end the long and barbaric reign of geopolitics and begin a new pilgrimage to create a lasting biosphere politics.

THERE ARE RARE MOMENTS in history when a generation of human beings are given a new gift with which to rearrange their relationship to one another and the world around them. This is such a moment. We are being given the power of the sun. Hydrogen is a promissory note for humanity's future on Earth. Whether that promise is squandered in failed ventures and lost opportunities or used wisely on behalf of our species and our fellow creatures is up to us.

NOTES

BETWEEN REALITIES

1. Rifkin, Jeremy. *The End of Work*. New York: Tarcher/Putnam, 1995.
2. Rifkin, Jeremy. *The Biotech Century*. New York: Tarcher/Putnam, 1998.
3. Rifkin, Jeremy. *The Age of Access*. New York: Tarcher/Putnam, 2000.
4. Albritton, Daniel L. et al. "Summary for Policy Makers: Climate Change 2001: A Report of Working Group I of the Intergovernmental Panel on Climate Change." IPCC, 2001. p. 13. *www.earth.usgcrp.gov/ipcc/wg1spm.pdf*
5. Hoffman, Peter. *The Forever Fuel: The Story of Hydrogen*. Boulder, CO. Westview Press, 1981.

SLIDING DOWN HUBBERT'S BELL CURVE

1. Crooks, Ed and David Buchan. "Crude Calculation." *Financial Times*. March 16/17, 2002. p. 6.
2. Campbell, Colin J.; Laherrère, Jean H. "The End of Cheap Oil." *Scientific American*. March 1998. p. 80; Ivanhoe, L. F. "Get Ready for Another Oil Shock!" *The Futurist*. January/February 1997. p. 23; Hatfield, Craig B. "How Long Can Oil Supply Grow?" Hubbert Center Newsletter #97/4. August 21, 2001. *www.hubbert.mines.edu/news/v97n4/mkh-new5.html*; MacKenzie, James J. "Oil as a Finite Resource: When Is Global Production Likely to Peak?" March 2000. *www.wri.org/wri/climate/jm_oil_000.html*; Banks, Howard. "Cheap Oil: Enjoy It While It Lasts." *Forbes*. June 15, 1998. p. 86.
3. Campbell, Colin J. "A Guide to Determining the World's Endowment and Depletion of Oil." March 31, 1998. *www.oilcrisis.com/campbell/guide.htm*
4. "Petroleum: Origin of Crude Oil." Encyclopaedia Britannica online.
5. Campbell. "A Guide to Determining the World's Endowment and Depletion of Oil."
6. Personal correspondence with C. J. Campbell (February 27, 2002) and Jean Laherrère (February 22, 2002).
7. Ivanhoe, L. F. "Future World Oil Supplies: There Is a Finite Limit." *World Oil*. October 1995. *www.users.knsi.com/~tbender/ivanhoe.html*

8. Ibid; Youngquist, Walter. *GeoDestinies: The Inevitable Control of Earth Resources over Nations and Individuals*. Portland, OR: National Book Company, 1997. p. 167.
9. Laherrère, Jean H. "World Oil Reserves—Which Number to Believe?" *OPEC Bulletin*. February 1995, pp. 9–13.
10. Youngquist. *GeoDestinies*. p. 171.
11. Ibid.
12. Personal correspondence with C. J. Campbell, February 27, 2002. Oil Depletion Analysis Center of London (ODAC) model to be published 2002.
13. Ibid; *World Oil* and *Oil and Gas Journal* figures on U.S. reserves are essentially identical. *World Oil* cites U.S. reserves at 21.33, and *Oil and Gas Journal* (*OGJ*) puts U.S. reserves at 22.05. The two journals, however, publish only the official reserve figures provided to them by governments. In the case of Saudi Arabia, *World Oil* puts reserves at 263 billion while *OGJ* lists reserves at 259.25 billion. Both figures are significantly higher than Campbell's assessment. See "World Trends: Estimated Proven World Reserves, 2000 versus 1999." *World Oil*. August, 2001, Vol. 222, No. 8. See also *Oil and Gas Journal*. December 24, 2001. pp. 126–7.
14. Laherrère. "World Oil Reserves—Which Number to Believe?"
15. Personal correspondence with C. J. Campbell, February 27, 2002, ODAC model to be published 2002; "Long-Term World Oil Supply (A Resource Base/Production Path Analysis)." The Energy Information Administration (EIA) of the Department of Energy, 2001.
16. "Long-Term World Oil Supply (A Resource Base/Production Path Analysis)." EIA, 2001.
17. Campbell and Laherrère. "The End of Cheap Oil." p. 79.
18. Ibid. p. 80.
19. Ibid. p. 80.
20. Banks. "Cheap Oil: Enjoy It While It Lasts." p. 84.
21. Campbell and Laherrère. "The End of Cheap Oil." p. 80.
22. Campbell, C. J. "Peak Oil." Presentation at the Technical University of Clausthal. December 2000. *www.oilcrisis.com/de/lecture.html*
23. "Estimated Ultimately Recoverable (EUR) Oil." *Oil as a Finite Resource*. World Resources Institute. *www.wri.org/wri/climate/finitoil/oil-eur.html*; Mackenzie. "Oil As a Finite Resource."
24. Campbell and Laherrère. "The End of Cheap Oil." p. 80.
25. Ivanhoe, L. F. "Get Ready for Another Oil Shock!" p. 22.
26. Campbell and Laherrère. "The End of Cheap Oil." p. 80; Campbell, Colin J. "Peak Oil." Presentation at the Technical University of Clausthal. December 2000. *www.oil crisis.com/de/lecture.html*
27. Ivanhoe. "Get Ready for Another Oil Shock!" p. 20.
28. "Estimated Ultimately Recoverable (EUR) Oil." World Resources Institute.
29. Ibid.
30. Ivanhoe. "Future World Oil Supplies: There Is a Finite Limit." Campbell. "A Guide to Determining the World's Endowment and Depletion of Oil."; Personal correspondence with L. F. (Buzz) Ivanhoe (March 4, 2002), C. J. Campbell (March 15, 2002), and Jean Laherrère (February 22, 2002).
31. Campbell, C. J. "The Next Oil Price Shock." *Energy Exploration and Exploitation*. Vol. 13, No. 1. 1995. p. 36.

32. Masters, Charles D., et al. "World Petroleum Assessment and Analysis." Proceedings of the 14th World Petroleum Congress. John Wiley & Sons, 1994. p. 537; "Estimated Ultimately Recoverable (EUR) Oil." World Resources Institute.
33. Personal correspondence with Jean Laherrère, February 22, 2002.
34. Ibid. Other geologists have slightly different figures. For example, geologist John D. Edwards of the University of Colorado reports that there are 25 billion barrels of crude oil produced each year compared to 7 billion barrels of recoverable oil discovered (including deep water) each year, making the ratio of production to discovery 3½ barrels to 1. Personal correspondence with John Edwards, March 14, 2002.
35. "2002 to Mark Last Hurrah for North Sea Oil." *Petroleum Intelligence Weekly*. February 4, 2002. p. 4.
36. Campbell and Laherrère. "The End of Cheap Oil." p. 82.
37. Ivanhoe. "Future World Oil Supplies: There Is a Finite Limit."
38. Cleveland, Cutler J.; Kaufmann, Robert K. "Why the Bush Oil (Energy) Policy Will Fail." October 5, 2001. *www.oilanalytics.org*; "Potential Oil Production from the Coastal Plain of the Arctic National Wildlife Refuge: Updated Assessment." EIA. May 2000; "Arctic National Wildlife Refuge, 1002 Area, Petroleum Assessment, 1998, Including Economic Analysis." United States Geological Survey (USGS). April 2001.
39. Banks. "Cheap Oil: Enjoy It While It Lasts." p. 84.
40. Ibid.
41. "Total Midyear Population for the World: 1950–2050. International Data Base. U.S. Bureau of the Census. May 10, 2000.
42. Youngquist. *GeoDestinies*. p. 198.
43. *Fortune* quoted in Romm and Curtis. "Mideast Oil Forever?" The Atlantic Online. April 1996. According to the EIA, world demand for oil in the year 2000 was between 74.47 and 76.62 million barrels per day. See "World Oil Demand 1997–2001." EIA. *www.eia.doc.gov/emeu/ipsr/t24/txt*
44. Carr, Edward. "Energy." *The Economist*. June 18, 1994. pp. 3–18.
45. "World Oil Markets." *International Energy Outlook 2001*. Energy Information Administration; Richard Reese. "Oil and the Future." May 31, 1997. p. 26. *www.unipri.it/~deyoung/oil_and_the_future.html*; Borenstein, Seth. "Oil is Plentiful in the World, But It'll Cost Us to Find It." *Akron Beacon Journal*. September 26, 2000. *www.ohio.com/bj/business/2000/September/26/docs/028918.html*
46. Youngquist. *GeoDestinies*. p. 204; "Long Term World Oil Supply (A Resource Base/Production Path Analysis)." EIA. 2001.
47. "Estimated Ultimately Recoverable (EUR) Oil." World Resources Institute; U.S. Geological Survey World Petroleum. Assessment 2000—Description and Results." USGS World Energy Assessment Team.
48. Edwards, John D. "Twenty-First Century Energy: Transition from Fossil Fuels to Renewable, Non-polluting Energy Sources." University of Colorado, Department of Geological Sciences—EMARC. April 2001; personal correspondence with John Edwards, March 14, 2002.
49. Ibid.
50. MacKenzie, "Oil as a Finite Resource."
51. M. K. Hubbert. "The Energy Resources of the Earth." *Scientific American*. September 1971. pp. 60–70; for more information on the Hubbert Curve refer to the M. King

Hubbert Center for Petroleum Supply Studies at the Colorado School of Mines founded by L. F. (Buzz) Ivanhoe. *www.hubbert.mines.com*

52. Campbell and Laherrère. "The End of Cheap Oil." p. 80.
53. Youngquist. *GeoDestinies*. p. 203.
54. Campbell and Laherrère. "The End of Cheap Oil." p. 80.
55. Personal correspondence with L. F. (Buzz) Ivanhoe, March 4, 2002.
56. Ivanhoe. "Get Ready For Another Oil Shock!" p. 23.
57. Hatfield. "How Long Can Oil Supply Grow?"
58. Ibid.
59. MacKenzie. "Oil as a Finite Resource."
60. Banks. "Cheap Oil: Enjoy It While It Lasts." p. 84.
61. Deffeyes, Kenneth S. *Hubbert's Peak: The Impending World Oil Shortage*. Princeton, NJ: Princeton University Press, 2001. pp. 146, 149.
62. Ibid. p. 149.
63. Ibid.
64. "Carbon Emissons Set to Rise Steadily with Fossil Fuel Use; But IEA Indicates Ways to Halt Climate Change Do Exist." Press Release. International Energy Agency. November 21, 2000; "World Energy Outlook." International Energy Agency. 1998.
65. Anderson, Roger N. "Oil Production in the 21st Century" *Scientific American*. Vol. 278. No. 3. March 1998. pp. 86–91.; Lynch, Michael C. "Closed Coffin: Ending the Debate on 'The End of Cheap Oil.' A Commentary." *www.sepwww.stanford.edu/sep/jon/world-oil.dir/lynch2.html*
66. MacKenzie, "Oil As a Finite Resource."; "U.S. Geological Survey World Petroleum Assessment 2000—Description and Results." U.S. Geological Survey World Energy Assessment Team.
67. Campbell, C. J. "Prophet or Cassandra?" Interview. *Petroleum Economist*. October 1995.
68. Campbell and Laherrère. "The End of Cheap Oil." p. 81.
69. "Estimated Ultimately Recoverable (EUR) Oil." World Resources Institute.
70. Fisher quoted in Richard A. Kerr. "The Next Oil Crisis Looms Large—and Perhaps Close." *Science Magazine*. August 1998. pp. 1128–1131.
71. Lynch. "Closed Coffin."
72. Anderson. "Oil Production in the 21st Century." p. 87.
73. Ibid. p. 87.
74. Ibid. p. 88.
75. Campbell and Laherrère. "The End of Cheap Oil." p. 82; Anderson, "Oil Production in the 21st Century." p. 88.
76. Campbell quoted in Kerr. "The Next Oil Crisis Looms Large."
77. Bartlett quoted in Kerr. "The Next Oil Crisis Looms Large."
78. Anderson. "Oil Production in the 21st Century." pp. 86–91.
79. Youngquist quoted in Richard Reese. "Oil and the Future." May 31, 1997. *www.unipri. it/~deyoung/oil_and_the_future.htm*
80. Youngquist. *GeoDestinies*. p. 173.
81. Ibid. p. 172; "World Crude Oil and Natural Gas Reserves, January 1, 2000." EIA updated February 5, 2001. *www.eia.doe/gov/emeu/iea/tables81.html*; "World Oil Sup-

ply, 1990–Present." EIA. *www.eia.doe.gov/emeu/ipsr/t14.txt*; "World Oil Demand, 1997–2001." EIA. *www.eia.doe.gov/emeu/ipsr/t24.txt*

82. "Petroleum 1996: Issues and Trends." EIA. p. 68; Youngquist. *GeoDestinies*. p. 111; *Survey of Current Business*. Bureau of Economic Analysis. U.S. Department of Commerce. January 2001. p. 60.
83. Schwartz, Nelson D. and Sung, Jessica. "Breaking OPEC's Grip." *Fortune Magazine*. Dow Jones Publication Library. 2001.
84. Tavernise, Sabrina. "Russians to Keep Limits on Oil Exports Through June." *The New York Times*. March 21, 2002. p. W1.
85. Wines, Michael and Sabrina Tavernise. "Russian Oil Production Still Soars, for Better and Worse." *The New York Times*. Nov. 21, 2001. p. A3.
86. Ibid.
87. Youngquist. *GeoDestinies*. p. 188.
88. Ibid. p. 188.
89. Ibid. p. 180.
90. Ibid. p. 190.
91. Romm and Curtis. "Mideast Oil Forever?"; Youngquist. *GeoDestinies*. p. 203.
92. Campbell, Colin J. "Myth of Spare Capacity Setting the Stage for Another Oil Shock." *Oil and Gas Journal*. March 20, 2000. p. 21.
93. Campbell, C. J. "Depletion Patterns Show Change Due for Production of Conventional Oil." *Oil and Gas Journal*. Dec. 29, 1997. p. 37.
94. Youngquist. *GeoDestinies*. p. 189.
95. MacKenzie. "Oil as a Finite Resource."
96. Riva, Joseph P., Jr. "World Oil Production after Year 2000: Business as Usual or Crisis?" Congressional Research Service. Library of Congress. 95–925. August 1995.

ENERGY AND THE RISE AND FALL OF CIVILIZATIONS

1. White, Leslie A. *The Evolution of Culture*. New York: McGraw Hill Company, 1959. p. 33.
2. Odum, Howard T. *Environment, Power, and Society*. New York: Wiley-Interscience, 1971. p. 49.
3. Ibid. p. 26.
4. White, Leslie A. *The Science of Culture: A Study of Man and Civilization*. New York: Farrar, Straus, and Company, 1949. p. 371.
5. MacCurdy, George Grant. *Human Origins*: A Manual of Prehistory. New York: D. Appleton and Company, 1924.
6. White. *The Science of Culture*. p. 376.
7. Odum, Howard T. *Environment, Power, and Society*. p. 27.
8. White. *The Science of Culture*. p. 368.
9. Ibid. pp. 368–369.
10. Ibid. p. 374.
11. White. *The Science of Culture*. p. 369; White. *The Evolution of Culture*. pp. 41–42.
12. Ponting, Clive. *A Green History of the World: The Environment and the Collapse of Great Civilizations*. New York: Penguin Books, 1991. p. 269.

13. Ibid. p. 270.
14. Ibid.
15. Whitehead, Alfred North. *Science and the Modern World*. New York: New American Library, 1925. p. 50.
16. Einstein quoted in Miller, G. Tyler, Jr. *Energetics, Kinetics and Life*. Belmont, California: Wadsworth, 1971. p. 46.
17. Asimov, Isaac. "In the Game of Energy and Thermodynamics You Can't Even Break Even." *Smithsonian*. August 1970. p. 9.
18. Soddy, Frederick. *Matter and Energy*. Home University Series. London: Oxford University, 1912. p. 245.
19. Blum, Harold F. *Time's Arrow and Evolution*. Princeton, NJ: Princeton University Press, 1968. p. 94.
20. Schrödinger, Erwin. *What Is Life?* New York: Macmillan, 1947. pp. 72, 75.
21. White, Leslie A. "Tools, Techniques and Energy." In *Cultural and Social Anthropology*. D. Hammond, ed. New York: Macmillan, 1964. p. 28.
22. Miller. *Energetics, Kinetics and Life*. p. 291.
23. Ibid.
24. Lotka quoted in White. "Tools, Techniques and Energy." p. 28.
25. Lotka, Alfred J. "Contribution to the Energetics of Evolution." *Proceedings of the National Academy of Science*, 1922. 8:149; Lotka. "The Law of Evolution as a Marxian Principle." *Human Biology* 17. September 1945. p. 186.
26. *World Resources: A Guide to the Global Environment. A Report by World Resources Institute, Nation Environment Programme, United Nations Development Programme, and the World Bank*. Oxford University Press, 1996.
27. Youngquist, Walter. *GeoDestinies: The Inevitable Control of Earth Resources over Nations and Individuals*. Portland, OR: National Book Company. 1997. p. 22.
28. Ibid.
29. Ibid.
30. Ibid. p. 32.
31. *World Resources*. A Report by World Resources Institute, UNEP, UNDP and the World Bank; "Climate Change and Energy." World Resources Institute. February 2002. "Individual Emissions" EPA Global Warming Site: Trends—Temperature. *www.epa.gov/globalwarming/emissions/individual/index.html*
32. "Renewing America's Infrastructure: A Citizen's Guide." American Society of Civil Engineers, 2001. pp. 3, 6–7.
33. Locke, John. "Second Treatise." In *John Locke, Two Treatises of Government*. Peter Laslett, ed. Cambridge: Cambridge University Press, 1967. p. 315.
34. Ibid. p. 312.
35. Melko, Matthew. *The Nature of Civilizations*. Boston: Porter Sargent, 1969. pp. 16–17.
36. Spengler, Oswald. *The Decline of the West*. Translated by Charles Francis Atkinson. New York: Modern Library, 1962. pp. 16–17.
37. Ibid. p. 73.
38. Tainter, Joseph A. *The Collapse of Complex Societies*. Cambridge: Cambridge University Press, 1988. pp. 79–80.
39. Coulborn, Rushton. "Structure and Process in the Rise and Fall of Civilized Society." *Comparative Studies in Society and History*. Vol. 8., 1966. p. 415.

40. Tainter. *The Collapse of Complex Societies.*
41. Levy, Jean-Philippe. *The Economic Life of the Ancient World.* Translated by John G. Biram. Chicago: University of Chicago Press, 1967. pp. 62–65;
42. Jones, A. H. M. *The Later Roman Empire, 284–602: A Social, Economic and Administrative Survey.* Norman: University of Oklahoma Press, 1964. pp. 114–115. Frank, Tenney. *An Economic Survey of Ancient Rome*, Vol. V: *Rome and Italy of the Empire.* Baltimore: Johns Hopkins Press, 1940. pp. 7–9.
43. Gibbon, Edward. *The Decline and Fall of the Roman Empire.* New York: Modern Library, 1776–88. p. 142; Tenney. *An Economic Survey of Ancient Rome.* p. 7.
44. Levy, Jean-Philippe. *The Economic Life of the Ancient World.* pp. 69, 77.
45. Jones, A. H. M. *The Roman Economy: Studies in Ancient Economic and Administrative History.* Oxford: Basil Blackwell, 1974. pp. 116, 127.; Hammond, Mason. "Economic Stagnation in the Early Roman Empire." *Journal of Economic History, Supplement.* Vol. 6. pp. 75–76.
46. Tainter. *The Collapse of Complex Societies.* p. 133.
47. Ibid. p. 142.
48. Ibid. p. 145.
49. Simkhovitch, Vladimir G. "Rome's Fall Reconsidered." *Political Science Quarterly.* Vol. 23, No. 2, June 1916. p. 226.
50. McNeill, William H. *Plagues and Peoples.* Garden City, NY: Anchor/Doubleday, 1976. p. 116; Hughes, Donald J. *Ecology in Ancient Civilizations.* Albuquerque: University of New Mexico Press, 1975. p. 131.
51. Tainter. *The Collapse of Complex Societies.* p. 144.
52. Ibid.
53. Simkhovitch. "Rome's Fall Reconsidered." p. 237.
54. Tainter. *The Collapse of Complex Societies.* p. 150.
55. Ibid.
56. "Ancient Rome." *Encarta Online Encyclopedia*, 2001; Kenneth Harl. "Early Medieval and Byzantine Civilization: Constantine to Crusades." *www.tulane.edu/~august/H303/handouts/Population.html*

THE FOSSIL-FUEL ERA

1. Miller, G. Tyler, Jr. *Living in the Environment.* Willard, OH: Brooks/Cole Publishing Co., 2001. MacKenzie, Jim H. and M. P. Walsh. *Driving Forces: Motor Vehicle Trends and Their Implications for Global Warming, Energy Strategies, and Transportation Planning.* World Resources Institute, 1990; Brown, Lester R. "Paving the Planet: Cars and Crops Competing for Land." *Worldwatch Issue Alert.* Worldwatch Institute, February 14, 2001.
2. "Number of U.S. Aircraft, Vehicles, Vessels, and other Conveyances." Bureau of Transportation Statistics of the U.S. Department of Transportation; "Regional Jet Aircraft." *Flight International.* October 1997; "Top 20 World Merchant Fleets by Country of Owner." The Maritime Administration of the Department of Transportation. February 8, 2002; Petursdottir, Gudrun, Olafur Hannibalsson, and Jeremy M. M. Turner. "Fatalities in Fisheries." *Safety at Sea As an Integral Part of Fisheries Management.* Food and Agriculture Organization of the United Nations, 2001.
3. MacKenzie, James J. "Oil as a Finite Resource: When Is Global Production Likely to Peak? March 2000. *www.wri.org/wri/climate/jm_oil_000.html.*

4. Ibid.
5. Youngquist, Walter. *GeoDestinies: The Inevitable Control of Earth Resources over Nations and Individuals*. Portland, OR: National Book Company. 1997. p. 197.
6. Schurr, Sam H. and Bruce C. Netschert. *Energy in the American Economy 1850–1975: An Economic Study of Its History and Prospects*. Baltimore: Johns Hopkins Press, 1960. p. 31.
7. Ponting, Clive. *A Green History of the World: The Environment and the Collapse of Great Civilizations*. New York: Penguin Books, 1991. pp. 287–288.
8. "World Consumption of Primary Energy by Selected Country Groups (Btu) 1990–1999," EIA. Updated February 1, 2001. *www.eia.doe.gov/emeu/iea/table18.html*
9. McNeill, William. *Plagues and People*. New York: Doubleday/Anchor Books, 1976. p. 147.
10. Mumford, Lewis. *Technics and Civilization*. New York: Harcourt, Brace, 1934. pp. 119–20.
11. Ibid. p. 120.
12. Rifkin, Jeremy. *Entropy: Into the Greenhouse World*. New York: Bantam Books, 1989. p. 88.
13. Howes, Edmund, ed. *Stow's Annals*. London, 1631, quoted in W. H. G. Armytage, *A Social History of Engineering*. London: 1961.
14. Rifkin. *Entropy*. p. 89.
15. Wilkinson, Richard. *Poverty and Progress*. New York: Praeger, 1973. pp. 90, 102.
16. Ponting, Clive. *A Green History of the World: The Environment and the Collapse of Great Civilizations*. p. 292.
17. Anderson, Robert. *Fundamentals of the Petroleum Industry*. Norman: University of Oklahoma Press, 1984. p. 2.
18. Ibid. pp. 9, 15, 19.
19. Ibid. p. 20.
20. Ibid.
21. Ibid. pp. 20–22, 29–30; Yergin, Daniel. *The Prize*. New York: Simon and Schuster, 1992. p. 210.
22. Rae, John B. "The Internal-Combustion Engine on Wheels." In *Technology in Western Civilization*. Vol. II. Melvin Kranzberg and Carroll W. Pursell, Jr., eds. London: Oxford University Press, 1967. p. 120.
23. Anderson. *Fundamentals of the Petroleum Industry*. p. 22.
24. Yergin. *The Prize*. p. 208.
25. Mowbray. A. Q. *Road to Ruin*. Philadelphia: Lippincott, 1969. p. 15.
26. Schneider, K. R. *Autokind v. Mankind*. New York: Schocken, 1972. p. 123.
27. "The Dramatic Story of Oil's Influence on the World." *Oregon Focus*. January 1993. pp. 10–11.
28. Churchill, Randolph S. *Winston S. Churchill*, Vol. 2, *Young Statesman, 1901–1917*. Boston: Houghton Mifflin, 1967. pp. 545–547.
29. Fisher, John Arbuthnot. *Fear God and Dread Nought: The Correspondence of Admiral of the Fleet Lord Fisher of Kilverstone*. 2 vols. Arthur J. Marder, ed. Cambridge: Harvard University Press, 1952. p. 404.
30. Churchill. *Churchill*, Vol. 2. p. 1964.
31. Hart, Basil Liddell. *A History of the World War, 1914–1918*. London: Faber and Faber, 1934.
32. Yergin. *The Prize*. p. 183.

33. Nuremburg Military Tribunals. *Trials of War Criminals*, vol. 7. Washington, D.C.: GPO, 1953. pp. 793–803.
34. Yergin. *The Prize*. p. 333.
35. U.S. Strategic Bombing Survey. Oil Division. *Final Report*. 2nd ed. Washington, DC: USSBS, 1947.
36. Economides, Michael and Ronald Oligney. *The Color of Oil: The History, the Money and the Politics of the World's Biggest Business*. Katy, Texas: Round Oak Publishing, 2000. p. 75.
37. Yergin. *The Prize*. p. 337.
38. Economides. *The Color of Oil*. p. 76; Yergin. *The Prize*. p. 877.
39. Youngquist. *GeoDestinies*. p. 48.
40. Economides. *The Color of Oil*. p. 100.
41. Anderson. *Fundamentals of the Petroleum Industry*. p. ix.
42. Zimmerman, Ann. "Wal-Mart Stores Profit Rises 92% as Consumers Shop for Bargains." *The Wall Street Journal*. February 20, 2002.
43. "Ranking the World's Top Oil Companies." Energy Intelligence Group, 2001. p. 5.
44. Slocum, Tyson. "No Competition: Oil Industry Mergers Provide Higher Profits, Leave Consumers with Fewer Choices." Washington, D.C.: Public Citizen's Critical Mass Energy and Environment Program, May 31, 2001. p. i.
45. Slocum. "No Competition."
46. Ibid.
47. Ibid.
48. Economides. *The Color of Oil*. p. 83.
49. Ibid. p. 84.
50. Ibid. p. 85.
51. Ibid. p. 86.
52. Ibid. p. 89.
53. Anderson. *Fundamentals of the Petroleum Industry*. p. 271.
54. Youngquist. *GeoDestinies*. p. 165–166.
55. Anderson. *Fundamentals of the Petroleum Industry*. pp. 271–277.
56. Ibid. pp. 279, 286, 289.
57. Chandler, Alfred D., Jr. *The Visible Hand*. Cambridge: Belknap Press of Harvard University Press, 1977. p. 83.
58. Ibid. p. 86.
59. Legerbott, Stanley. "United States Transport and Externalities." *Journal of Economic History*: Vol. 26, December 1966. pp. 444–446.
60. Chandler, *The Visible Hand*. p. 91.
61. Ibid. pp. 107–109.
62. Ibid. p. 120.
63. Ibid. pp. 204–205.
64. Ibid. p. 89.
65. Kern, Stephen. *The Culture of Time and Space* 1880–1918. Cambridge: Harvard University Press, 1983. p. 12.
66. Landes, David S. *Revolution in Time*. Cambridge: Harvard University Press, 1983. pp. 285–286.
67. Taylor, Frederick. *The Principles of Scientific Management*. New York: W. W. Norton, 1947. pp. 235–236.

68. Ibid. pp. 39, 63.
69. Bell, Daniel. "The Clock Watchers: Americans at Work." *Time*. September 8, 1975. p. 55.
70. "Corporate Mergers." Federal Trade Commission. 1963. p. 176.
71. Renner, Michael. "Corporate Mergers Skyrocket." *Vital Signs*. Washington, D.C.: Worldwatch Institute, 2000. *www.globalpolicy.org*
72. Hansell, Saul. "America Online Agrees to Buy Time Warner for $165 Billion; Media Deal is Richest Merger." *The New York Times*. January 1, 2000. p. A1.
73. Renner. "Corporate Mergers Skyrocket."
74. Anderson, Sarah and John Cavanagh. "Top 200: A Profile of Global Corporate Power." Washington, D.C.: Interhemispheric Resource Center, 1996. p. 3.

THE ISLAMIST WILD CARD

1. Pipes, Daniel. *In the Path of God: Islam and Political Power*. New York: Basic Books, Inc. 1983. p. 288.
2. Banerjee, Neela. "World Oil at a Glance." *The New York Times*. October 14, 2001. p. C1.
3. *Britannica Book of the Year*. Chicago: Encyclopaedia Britannica, Inc., 1999; "The World Factbook 2001: "Central Intelligence Agency (CIA)." Facts About Islam." Harvard Islamic Society. *www.digitas.harvard.edu/~facts.html*
4. Wolf, Martin. "The Economic Failure of Islam." *Financial Times*. September 26, 2001. p. 17.
5. Ibid.
6. Quoted in Pipes. *In the Path of God*. p. 281.
7. Overbye, Dennis. "Islamic Science: Rise and Fall." *International Herald Tribune*. November 1, 2001. p. 8.
8. Ibid.
9. Quoted in Randall, John Herman. *The Making of the Modern Mind*. Boston: Houghton Mifflin, 1940. p. 223.
10. Bacon, Francis. *Novum Organum*. Book I. Aphorism 71.
11. Bacon, Francis. "Novum Organum." in *The Works of Francis Bacon*, Vol. 4. London: J. Rivington and Sons, 1778. pp. 114, 246, 320, 325.
12. Condorcet, Marie, Marquis de. *Outlines of an Historical View of the Progress of the Human Mind*. London: J. Johnson, 1795. pp. 4–5.
13. Davidson, Lawrence. *Islamic Fundamentalism*. Westport, Connecticut: Greenwood Press, 1998. pp. 12–13.
14. Pipes. *In the Path of God*. p. 285.
15. "Oil Price History and Analysis." *Energy Economics Newsletter*. WTRG Economics. *www.wtrg.con/prices.html*; Pipes. *In the Path of God*. p. 290.
16. Pipes. *In the Path of God*. p. 290.
17. Ibid. p. 290.
18. Ibid. p. 294.
19. Yergin, Daniel. *The Prize: The Epic Quest for Oil, Money and Power*. New York: Simon & Schuster, 1992. pp. 605–606.
20. Ibid. p. 608.
21. Ibid. p. 614.
22. Ibid.

23. Ibid. p. 625.
24. Ibid. p. 628.
25. Ibid.
26. Ibid. p. 635.
27. Steinbach, Udo. "Saudi Arabiens neue Rolle im Nahen Osten." *Aussenpolitik* 25. 1974. p. 210.
28. Cordesman, Anthony H. "Geopolitics and Energy in the Middle East." September 1999. p. vi. *www.csis.org.mideast/reports/Meenergy.html*
29. Gause III, F. Gregory. *Oil Monarchies: Domestic and Security Challenges in the Arab Gulf States*. New York: Council on Foreign Relations Press. 1994. p. 45.
30. Ibid. pp. 58–59.
31. Ibid. p. 69.
32. Cordesman. "Geopolitics and Energy in the Middle East." p. 13.
33. Ibid.
34. Ibid.
35. Youngquist, Walter. *GeoDestinies: The Inevitable Control of Earth Resources Over Nations and Individuals*. Portland, OR: National Book Company. 1997. p. 119.
36. Ibid. pp. 133–135.
37. *International Financial Statistics*. CD-ROM version. International Monetary Fund. 2000; "Saudi Arabia." *The World Factbook*. CIA. 2001.
38. Cordesman. "Geopolitics and Energy in the Middle East." p. 28; "Saudi Arabia." *The World Factbook*. CIA.
39. Cordesman. "Geopolitics and Energy in the Middle East." p. 29.
40. Ibid. pp. 29–31; "Saudi Arabia." *The World Factbook*. CIA.
41. Cordesman. "Geopolitics and Energy in the Middle East." p. 31.
42. Ibid.
43. Ibid.
44. Ibid. pp. xi; 30.
45. Wolf. "The Economic Failure of Islam." p. 17.
46. Worth, Robert. "The Deep Intellectual Roots of Islamic Terror." *The New York Times*. October 13, 2001. p. A13.
47. Ajami quoted in Zakaria, Fareed. "Why Do They Hate Us?" *Newsweek*. October 15, 2001. p. 33.
48. Waldman, Peter; Pope, Hugh. "'Crusade' Reference Reinforces Fears on War on Terrorism Is Against Muslims." *The Wall Street Journal*. September 21, 2001. p. A1.
49. Jehl, Douglas. "A Saudi Prince with an Unconventional Idea: Elections." *The New York Times*. November 28, 2001. p. A3.
50. Armstrong, Karen. *Islam: A Short History*. New York: The Modern Library, 2000. p. 160.
51. Davidson, Lawrence. *Islamic Fundamentalism*. Westport, Connecticut: Greenwood Press, 1998. pp. 14–15.
52. Jehl, Douglas. "Democracy's Uneasy Steps in Islamic World." *The New York Times*. November 23, 2001. p. A1.
53. Ibid. p. A10.
54. Ibid.
55. Ashby, Tom. "World Oil Prices Soar on Iraq's Decision." *Reuters*. April 8, 2002;

Hafidh, Hassan. "Iraq Suspends Oil Exports to Support Palestinians." *Reuters*. April 8, 2002.

56. Fathi, Nazila. "Iranian Urges Muslims to Use Oil As a Weapon." *The New York Times*. April 6, 2002. p. A9; Hafidh. "Iraq Suspends Oil Exports to Support Palestinians."
57. Pincus, Ward. "Gulf Officials Say Oil Boycott of United States Not Practical." *The Associated Press*. April 3, 2002.
58. Ibid.
59. Ibid.
60. Romm, Joseph J.; Curtis, Charles B. "Mideast Oil Forever?" *The Atlantic Online*. April 1996. *www.theatlantic.com//issues/96apr/oil/oil.html*

A GLOBAL MELTDOWN

1. Ausubel, Jesse H. "Energy and Environment: The Light Path." *Energy Systems and Policy*. Vol. 15. 1991. *www.phe.rockefeller.edu/light_path/*
2. "Energy Needs, Choices and Possibilities: Scenarios to 2050." Shell Oil International, 2001. *www2.shell.com*
3. Cordesman, Anthony H. "Geopolitics and Energy in the Middle East." September 1999. p. 26. *www.csis.org/mideast/reports/Meenergy.html*
4. "Natural Gas Consumption in the United States, 1996–2002." *Natural Gas Monthly*. EIA, February 2002.
5. Personal correspondence with Richard Duncan, Director of The Institute on Energy and Man. March 18, 2002.
6. "Market Trends—Oil & Natural Gas." *Annual Energy Outlook 2001*. Report #: DOE/EIA-0383(2001) December 22, 2000. *www.eia.doe.gov/oiaf/aeo/gas.html*
7. Ibid.
8. Personal correspondence with Richard Duncan. March 18, 2002.
9. True, Warren R. "Major Gas Projects Fuel Surge in Long-Term Plans." *Oil and Gas Journal*. February 5, 2001.
10. Personal correspondence with Richard Duncan. March 18, 2002.
11. Ibid.
12. Ibid.
13. Duncan, Richard C. "The Peak of World Oil Production and the Road to the Olduvai Gorge" *Pardec Keynote Symposia*. Geological Society of America Summit 2000. Reno, Nevada. November 13, 2000.
14. Youngquist, Walter. *GeoDestinies: The Inevitable Control of Earth Resources over Nations and Individuals*. Portland, OR: National Book Company. 1997. p. 196.
15. Campbell, C. J. "Depletion Patterns Show Change Due for Production of Conventional Oil." *Oil and Gas Journal*. Dec. 29, 1997. p. 37.
16. Cordesman, Anthony H. "Geopolitics and Energy in the Middle East." September 1999. p. 26. *www.csis.org/mideast/reports/Meenergy.html*
17. "World Energy Outlook." International Energy Agency. 1998. *www.iea.org/new/releases/weo1.html*
18. Holdren, John P. "The Energy Climate Challenge: Issues for the New U.S. Administration." *Environment*. June 2001.
19. Ibid.

20. "Coal." *International Energy Outlook 2001*. March 2001. *www.eia.doe.gov/oiaf/ieo/index.html*
21. "The Energy Information Administration (EIA) Recently Updated the Energy." *Electrical World*. Vol. 215, No. 4, July, 2001. p. 14.
22. Ruth, Lawrence A. "Advanced Coal-Fired Power Plants." *Journal of Energy Resources Technology*. Vol. 123. March 2001. p. 4.
23. "Coal." *International Energy Outlook 2001*. March 2001. *www.eia.doe.gov/oiaf/ieo/index.html*
24. Dunne, Nancy. "King Coal Makes a Comeback" *Financial Times*. August 8, 2001.
25. Ibid.
26. Hatfield, Craig B. "How Long Can Oil Supply Grow?" Hubbert Center Newsletter #97/4. August 21, 2001. *www.hubbertmines.edu/news/v97n4/mkh-new5.html*
27. Ibid.
28. Lanier, Douglas. "Heavy Oil—A Major Energy Source for the 21st Century." UNITAR Centre for Heavy Crude and Tar Sands. No. 1998.039. 1998. p. 1.; George, Richard L. "Mining for Oil." *Scientific American*. Vol. 278. No. 3. March 1998. pp. 84–85.; Masters, Charles D.; Emil D. Attanasi; David H. Root. "World Petroleum Assessment and Analysis." US DOE. World Petroleum Conference. October 5, 2001. *www.energy.er.usgs.gov/products/papers/WPC/14/text.htm*
29. George. "Mining for Oil." pp. 84–85.
30. Lanier, Douglas. "Heavy Oil—A Major Energy Source for the 21st Century." UNITAR Centre for Heavy Crude and Tar Sands. No. 1998.039. 1998. p. 2.
31. Stosur, George J. et al. "Tar Sands—Technology, Economics, and Environmental Issues for Commercial Production Beyond Year 2000." United Nations Institute for Training and Research—International Centre for Heavy Hydrocarbons. No. 1998.002. p. 1.
32. Lanier, Douglas. "Heavy Oil—A Major Energy Source for the 21st Century." pp. 1–2.
33. Moritis, Guntis. "Massive Oil Resource to Be Targeted by New EOR Techniques." *Oil and Gas Journal*. December 13, 1999. *www.findarticles.com/cf_0m3112/50_97/58500622/print.jhtml*
34. Lanier, Douglas. "Heavy Oil—A Major Energy Source for the 21st Century." p. 2.
35. Haaland, O. R. Klovning and T. Sem. "The Future of the World's Extra Heavy Oil Resources—Competition and Potential." UNITAR Centre for Heavy Crude and Tar Sands. No. 1998.008. 1998. p. 4.
36. "What We Do at Syncrude." Syncrude. 2001. *www.syncrude.com/who_we_are/01_04.html*
37. "Shell Canada, Chevron Begins Oilsands Project" Business and Industry. Vol. 15, No. 1; p. 6; ISSN: 0009–272X. January 2000.
38. Schwartz, Nelson D. and Jessica Sung. "Breaking OPEC's Grip." *Fortune Magazine*. Dow Jones Publication Library. 2001; Lanier, Douglas. "Heavy Oil—A Major Energy Source for the 21st Century" UNITAR Centre for Heavy Crude and Tar Sands. No. 1998.039. 1998. p. 2.
39. Schwartz, Nelson D. and Jessica Sung. "Breaking OPEC's Grip." *Fortune Magazine*. Dow Jones Publication Library. 2001.
40. Ibid; Stosur, George J. et al. "Tar Sands—Technology, Economics, and Environmental Issues for Commercial Production Beyond Year 2000." p. 4.
41. Schwartz, Nelson D. and Jessica Sung. "Breaking OPEC's Grip."

42. Personal correspondence with Jean Laherrère, March 4, 2002; Nikiforuk, A. "The Next Gas Crisis." *Canadian Business*. August 20, 2001.
43. Stosur, George J.; et al. "Tar Sands—Technology, Economics, and Environmental Issues for Commercial Production Beyond Year 2000." p. 4.
44. Sapre, Alex et al. "Synthetic Fuels, Carbon Dioxide and Climate." *Responsible Interpretation of Atmospheric Models and Related Data*. R.A. Reck and J. R. Hummel, eds. American Institute of Physics Press, 1982. pp. 135, 140.
45. Stosur, George J. et al. "Tar Sands—Technology, Economics, and Environmental Issues for Commercial Production Beyond Year 2000." p. 4.
46. Albritton, Daniel L. et al. "Summary for Policy Makers: Climate Change 2001: A Report of Working Group I of the Intergovernmental Panel on Climate Change." IPCC, 2001. p. 7. *www.earth.usgcrp.gov/ipcc/wg1spm.pdf*
47. Ibid.
48. Ibid.
49. Houghton, John. *Global Warming: The Complete Briefing*. Cambridge: Cambridge University Press. 1997. p. 22.
50. Ibid. p. 12.
51. Albritton. "Summary for Policy Makers: Climate Change 2001." p. 2.
52. Ibid. p. 13.
53. Ibid.
54. Anonymous. "Global Warming: Fact vs. Myth." *Environmental Defense. www.environmentaldefense.org/pubs/FactSheet/e_GWFact2.html*
55. Lemonick, Michael D. "Life in the Greenhouse." *Time*. April 9, 2001. p. 27.
56. Ibid.
57. Ibid. p. 25.
58. Ibid. p. 27.
59. Ibid.
60. Monastersky, Richard. "The Long Goodbye." *New Scientist*. No. 2286. April 14, 2001. p. 31.
61. Albritton. "Summary for Policy Makers: Climate Change 2001." p. 4.
62. Ibid.
63. Ibid. pp. 4, 16.
64. Pearce, Fred. "Washed off the Map." *New Scientist*. No. 2266. November 25, 2000. p. 5.
65. Ibid.
66. Ibid.
67. Ibid; Lemonick, Michael D. "Life in the Greenhouse." *Time*. April 9, 2001. p. 29.
68. Houghton. *Global Warming: The Complete Briefing*. p. 114.
69. Ibid. pp. 11, 13.
70. Albritton. "Summary for Policy Makers: Climate Change 2001." p. 4.
71. Ibid; "Wild Weather." *New Scientist*. September 16, 2000. p. 29.
72. Ahmad, Q.K. et al. "Summary for Policy Makers, Climate Change 2001: Impacts, Adaptation, and Vulnerability." A Report of Working Group II of the Intergovernmental Panel on Climate Change. IPCC. February, 2001. p. 13.
73. Hurtt, G. C. and S. Hale. "Future Climates of the New England Region." In *Preparing for a Changing Climate: The New England Regional Assessment Overview* (by New England Regional Assessment Group). U.S. Global Change Research Program. University of New Hampshire, 2001.

74. Houghton. *Global Warming: The Complete Briefing*. p. 127.
75. Ibid.
76. Ibid.
77. Clark quoted in Beardsley, Tim. "In the Heat of the Night." *Scientific American*. October 1998. p. 20.
78. Ibid.
79. Cox cited in Pearce, Fred. "Violent Future." *New Scientist*. July 21, 2001. p. 4.
80. Pearce, Fred. "Global Green Belt." *New Scientist*. September 15, 2001. p. 15.
81. Ahmad, "Summary for Policy Makers, Climate Change 2001: Impacts, Adaptation, and Vulnerability." p. 11.
82. Connor, Steve. "British Association: Coral Reefs Will All Die Within 50 Years, Study Finds Constantly Rising Sea Temperatures Are Causing the Bleaching and Death of Fragile Sea Structures, Warns Marine Specialist." *Financial Times*. September 6, 2001.
83. Pearce, Fred. "A Searing Future." *New Scientist*. No. 2264. November 11, 2000. p. 4.
84. Lemonick, Michael D. "Life in the Greenhouse." *Time*. April 9, 2001. p. 28.
85. Ahmad, "Summary for Policy Makers, Climate Change 2001: Impacts, Adaptation, and Vulnerability." p. 8.
86. Ibid. p. 12.
87. Houghton, John. *Global Warming: The Complete Briefing*. Cambridge: Cambridge University Press. 1997. p. 132.
88. Alley, R. B. "The Younger Dryas Cold Interval As Viewed From Central Greenland." *Quaternary Science Reviews*. Vol. 19, 2000. pp. 213–226.
89. "Abrupt Climate Change: Inevitable Surprises." The National Academy of Sciences. Washington, D.C.: National Academy Press, 2002. p. 10.
90. Ibid. p. 118.
91. Ibid. p. 92.
92. Ibid. p. 86.
93. Peteet, D. M. et al. "Late-glacial Pollen, Macrofossils, and Fish Remain in Northeastern USA—The Younger Dryas Oscillation." *Quaternary Science Reviews*. Vol. 12, 1993. pp. 597–612; Martin, P. S. "Prehistoric Overkill: The Global Model." In P. S. Martin and R. G. Klein, eds., *Quaternary Extinctions*. Tucson, AZ: University of Arizona Press, 1984. pp. 354–403.
94. Overpeck, J. "Warm Climate Surprises." *Science*. Vol. 271, 1996. pp. 1820–1821; "Abrupt Climate Change: Inevitable Surprises." The National Academy of Sciences. p. 113.
95. "Abrupt Climate Change: Inevitable Surprises." The National Academy of Sciences. pp. 88, 119.

VULNERABILITIES ALONG THE SEAMS

1. "Gulf War Facts." CNN News Online. *www.cnn.com/SPECIALS/2001/gulf.war/facts/gulfwar/index.html*
2. Ibid.: "Conduct of the Persian Gulf War." The Final Report to the U.S. Congress by the U.S. Department of Defense. April 1992. Appendix P.
3. Narrator. "Fighting for Oil." Show Transcript. America's Defense Monitor. Center for Defense Information. Washington, D.C.: January 28, 1996.

4. Telhami, Shibley and Joseph J. Romm. "Fighting for Oil." Show Transcript. America's Defense Monitor. Center for Defense Information. Washington, D.C.: January 28, 1996.
5. Cortese, Amy. "New York's Setback Expected to Be Deeper than Nation's." *The New York Times*. December 17, 2001. p. C4.
6. Ibid.
7. Wax, Alan J. "Unemployment Rate in NYC Hits 7.4%." January 18, 2002. *www.newsday.com*
8. Baily, Martin Neil. "Economic Policy Following the Terrorist Attack." Institute for International Economics. *www.icc.com/policybriefs/news01-10.html*
9. "Attacks May Cost U.S. 1.8 Million Jobs." *The New York Times*. January 13, 2002. p. A16.
10. "Decrease in Tourism Demand Signals the Loss of Millions of Jobs Worldwide." World Travel and Tourism Council. London: September 25, 2001.
11. Ibid.
12. "Declaration of Global Travel and Tourism Associations." October 2, 2001.
13. "Decrease in Tourism Demand Signals the Loss of Millions of Jobs Worldwide." World Travel and Tourism Council.
14. Allen, Mike and Amy Goldstein. "Security Funding Tops New Budget." *The Washington Post*. January 20, 2002. p. A1.
15. Purdum, Todd and Howard W. French. "U.S. Makes Pledge for $300 Million in Aid to Afghans." *The New York Times*. January 21, 2002. p. A1.
16. Sun, Lena H. and Jacqueline L. Salmon. "U.S. Sets Formula to Pay Victims." *The Washington Post*. December 21, 2001. p. A1; Hernandez, Raymond. "Negotiators Back $8.2 Billion in Aid for New York City." *The New York Times*. December 19, 2001. p. A1.; Alvarez, Lizette and Stephen Labaton. "A Nation Challenged: The Bailout; An Airline Bailout." *The New York Times*. September 22, 2001. p. A1.
17. "President Establishes Office of Homeland Security." The White House. Washington, D.C. *www.whitehouse.gov/news/releases/2001/10/print/20011008.html*; "Homeland Security and the President's Budget Priorities." The White House. Washington, D.C. *www.whitehouse.gov/homeland*
18. "Lawmakers Approve Anti-Terrorism Funds Package." Reuters. December 18, 2001.
19. Mitchell, Alison. "A Nation Challenged: The Domestic Front; Ridge Promises Security Funds for States in Next U.S. Budget." *The New York Times*. December 7, 2001. p. B7.
20. Schoffner, Chuck. "Microbiologist Favors Flyover Bans." *The Los Angeles Times*. September 21, 2001.
21. Horrock, Nicholas. "The New Terror Fear—Biological Weapons: Detecting an Attack Is Just the First Problem." *U.S. News & World Report*. May 12, 1997. p. 36.
22. Ibid.
23. Ibid.
24. Cole, Leonard A. "The Specter of Biological Weapons." *Scientific American*. December 1996. p. 62.
25. Gever, John. *Beyond Oil*. Denver, Co: University Press of Colorado, 1991. p. 172.
26. Goodman, David, et al. *From Farming to Biotechnology: A Theory of Agro-Industrial Development*. New York: Basil Blackwell, 1987. p. 25.

27. Cochrane, Willard. *Development of American Agriculture: A Historical Analysis*. 2nd ed. Minneapolis: University of Minnesota Press, 1993. p. 126.
28. Ibid., p. 197.
29. Brown, Lester R. et al. *State of the World 1990*. Washington, D.C.: Worldwatch Institute; New York: W.W. Norton, 1990. p. 67.
30. "Pesticide Industry Sales and Usage: 1986 Market Estimates." Economic Analysis Branch, Benefits and Use Division, Office of Pesticides Programs, Environmental Protection Agency. August 1987.
31. "Why Job Growth Is Stalled." *Fortune*. March 8, 1993. p. 52.
32. "The Mechanization of Agriculture." *Scientific American*. September 1982. p. 77.
33. Cochrane. *Development of American Agriculture*. pp. 137, 158–159.
34. Farb, Peter. *Humankind*. Boston: Houghton Mifflin, 1978. pp. 181–182.
35. Rifkin, Jeremy. *Entropy: Into the Greenhouse World*. Rev. ed. New York: Bantam Books, 1980, 1989. p. 155.
36. *Environmental Quality*. Ninth Annual Report of the Council on Environmental Quality. Washington, D.C.: U.S. Government Printing Office, December 1978. p. 270.
37. Clark, Wilson. *Energy for Survival*. Garden City, NY: Doubleday/Anchor Books, 1975. p. 170.
38. Ponting, Clive. *A Green History of the World: The Environment and the Collapse of Great Civilizations*. New York: Penguin Books, 1991. p. 292.
39. Reimund, Donn A. and Judith Z. Kalbacher. *Characteristics of Large-Scale Farms, 1987*. Washington, DC: USDA Economic Research Service, April 1993. p. iii.
40. "The Developing World's New Burden: Obesity." Food and Agriculture Organization of the United Nations. January 2002. *www.fao.org/FOCUS/E/obesity/obes1.htm*
41. Ponting. *A Green History of the World*. p. 404.
42. Halweil, Brian. "Worldwatch Press Briefing on the Global Trends in Meat Consumption." Worldwatch Institute. July 2, 1998. *www.worldwatch.org/alerts*; Durning, Alan B. "Cost of Beef for Health and Habitat." *The Los Angeles Times*. September 21, 1986. p. 3.
43. Based on sixty-five pounds of beef consumed per person per year. The auto CO_2 emissions comparisons come from Andrew Kimbrell, "On the Road," In Jeremy Rifkin, ed., *The Green Lifestyle Handbook*. New York: Owl Books, 1990.
44. Based on data from David Pimentel. *Food, Energy and Society*. College of Agriculture and Life Sciences and Division of Nutritional Sciences. Cornell University. Ithaca, NY. *www.unu.edu/unupress/food/8F072e/8F072E06.htm*. Statistics calculated by Steve Morningthunder of Instituto de Física, Universidad Nacional Autónoma de Mexico.
45. Ponting. *A Green History of the World*. p. 240.
46. Adams, Richard Newbold. *Energy and Structure: A Theory of Social Power*. Austin, Texas: University of Texas Press, 1975. p. 266.
47. Sale, Kirkpatrick. "The Polis Perplexity: An Inquiry into the Size of Cities." Working Papers, January–February 1978. p. 66.; Barbara Ward. *The Home of Man*. New York: Norton, 1976. p. 4.
48. "Cities at the Forefront." *Population Reports*. Johns Hopkins University. 2001. *www.jhuccp.org/pr/urbanpre.stm*
49. Wolman, A. "The Metabolism of Cities." *Scientific American*. No. 213, 1965. pp. 178–190.

50. Schwartz, John. "Securing the Lines of a Wired Nation." *The New York Times*. October 4, 2001. p. F8.
51. Ibid.
52. "'65 Northeast Power Loss Was Shorter, But Bigger." *The Washington Post*. July 15, 1977. p. A10.
53. Ibid.; Haberman, Clyde. "'77 Blackout: The Heart of Darkness." *The New York Times*. July 11, 1997. p. B1.
54. Connell, Rich. "Massive Power Outage Hits 7 Western States." *The Los Angeles Times*. August 11, 1996 p. A1.; Martha L. Willman. "Customers Will Foot Bill for Outage Repairs." *The Los Angeles Times*. October 19, 1996. p. D1.
55. Booth, William. "800,000 Lose Power in California As Blackouts Roll Across the State." *The Washington Post*. March 20, 2001. p. A8.
56. Huber, Peter W. "Dig More Coal—The PCs Are Coming." *Forbes Magazine*. May 31, 1999. p. 70.
57. Ibid.
58. Ibid.
59. Ibid. p. 71.
60. Lovins, Amory B. and L. Hunter Lovins. *Brittle Power: Energy Strategy for National Security*. Rocky Mountain Institute Andover, MA: Brick House Publishing Co., Inc., 2002. p. 35.
61. Ibid. pp. 36–37.
62. Ibid. p. 39.
63. Ibid.
64. Stephens, M. M. "The Oil and Natural Gas Industries: A Potential Target of Terrorists." In *Terrorism: Threat, Reality, Response*. R. Kupperman and D. Trent, eds. Stanford University: Hoover Institution Press, 1980. p. 206; M. M. Stephens. *Vulnerability of Total Petroleum Systems*. May report to Defense Civil Preparedness Agency. Washington, D.C. #DAHC20-70-C-0315, DCPA Work Unit 4362A, 1973.
65. Lovins and Lovins. *Brittle Power*. pp. 46–47; Stephens. *Vulnerability of Total Petroleum Systems*.
66. Lovins and Lovins. *Brittle Power*. p. 59.
67. Stephens. *Vulnerability of Total Petroleum Systems*. p. iv.
68. *National Energy Transportation*. Congressional Research Service. Vol. III. Report to U.S. Senate Committees on Energy and Natural Resources and on Commerce, Science and Transportation. May 1977. USGPO #95–15. pp. 159–160.
69. *Civil Preparedness Review*. Joint Committee on Defense Production. Joint Committee Print. U.S. Congress, 95th Congress, 1st Session. February 1977. USGPO. Vol. A. p. 1.
70. Lovins and Lovins. *Brittle Power*. p. 127.
71. Beyea, J. "Some Long-Term Consequences of Hypothetical Major Releases of Radioactivity to the Atmosphere from Three-Mile Island." Report PU/CEES #109 to President's Council on Environmental Quality. Princeton, NJ: Center for Energy and Environmental Studies, Princeton University. 1980.
72. "Protection of Electric Power Systems." Defense Electric Power Administration. Research Project 4405. Washington, D.C.: Department of the Interior, June 1962. pp. 25–26.
73. *Civil Preparedness Review*. Joint Committee on Defense Production. 1977. p. 87.

74. BP Amoco statistical review of world energy. London: BP Amoco, 2000. pp. 11, 40. *www.bpamoco.com/worldenergy*

THE DAWN OF THE HYDROGEN ECONOMY

1. Verne, Jules. *The Mysterious Island*. Translated by W. H. G. Kingston. New York: The Limited Editions Club at the Garamond Press, 1959.
2. Moore, Bill. "The Day the World Came to Its Senses?" *EV World*. October 12, 2001. *www.evworld.com/databases/storybuilder.cfm?storyid=245*
3. "Hydrogen." *The Columbia Encyclopedia*, Sixth Edition. Columbia University Press, 2001.
4. Nakicenovic, Nebojsa. "Freeing Energy from Carbon." *Daedalus*. Vol. 125, No. 3. Summer 1996. pp. 98–99.
5. Ausubel, Jesse H. "Where Is the Energy Going?" *The Industrial Physicist*. February 2000.
6. Hydrogen Technical Advisory Panel (HTAP). US DOE. "Fuel Choice for Fuel Cell Vehicles." Washington, D.C.: May 1999.
7. Ingriselli, Frank. "Powering Future Mobility with Electric Transportation Technologies." Presentation to House Science Committee. U.S. House of Representatives. April 23, 2001.
8. HTAP. "Fuel Choice for Fuel Cell Vehicles."
9. Hoffmann, Peter. *Tomorrow's Energy: Hydrogen, Fuel Cells, and the Prospects for a Cleaner Planet*. Cambridge, MA: The MIT Press, 2001. pp. 22–23.
10. Ibid. p. 23.
11. Ibid. pp. 23–24.
12. Ibid. pp. 29–30.
13. Haldane, J. B. S. *Daedalus or Science and the Future*. Dutton: 1925.
14. Ibid.
15. Ibid.
16. Ibid.
17. Hoffmann. *Tomorrow's Energy*. p. 32.
18. Dunn, Seth. "Hydrogen Futures: Toward a Sustainable Energy System." Worldwatch Paper 157. Washington, D.C.: Worldwatch Institute, August 2001. p. 28.
19. Hoffmann. *Tomorrow's Energy*. pp. 212–213.
20. Ibid. p. 42.
21. Hoffmann. *Tomorrow's Energy*. p. 42.
22. Ibid. pp. 46–50, 247.
23. Ibid. p. 247.
24. Pearce, Fred. "Kicking the Habit." *New Scientist Magazine*. Vol. 168, No. 2266. November 25, 2000. p. 34.
25. Dunn. "Hydrogen Futures."; Koppel, Tom. "Renewable Energy in the Island State." *Refocus*. June 2001. p. 5; Carl T. Hall. "Hydrogen Powers Energy Hopes: Experts Say It May Be the Fuel of the Future." *San Francisco Chronicle*. April 2, 2001.
26. Ogden, Joan M. "Prospects for Building a Hydrogen Energy Infrastructure." *Annual Review of Energy and Environment*. Vol. 24. 1999. pp. 227–279.
27. Ibid.
28. Dunn. "Hydrogen Futures." p. 31.

29. Chambers, Ann. *Distributed Generation: A Nontechnical Guide*. Tulsa, OK: PennWell. 2001. p. 150.
30. Ibid.
31. Hoffmann. *Tomorrow's Energy*. pp. 59–60.
32. Padro, C.E.G. and V. Putsche. "Survey of the Economics of Hydrogen Technologies." Technical Report. Golden, CO: NREL. September 1999.
33. Institute of Gas Technology. "Survey of Hydrogen Production and Utilization Methods." 1975.
34. Dunn. "Hydrogen Futures." p. 32.
35. Houghton, John. *Global Warming: The Complete Briefing*. Cambridge, MA: Cambridge University Press, 1997. p. 203.
36. Hawken, Paul, Amory Lovins, and Hunter L. Lovins. *Natural Capitalism: Creating the Next Industrial Revolution*. Boston: Little, Brown and Company, 1999. p. 248.
37. Houghton. *Global Warming*. pp. 210–211.
38. Hoffmann. *Tomorrow's Energy*. p. 90.
39. Silberman, Steve. "The Energy Web." *Wired*. July 2001. p. 119.
40. Borberly, Anne-Marie and Jan F. Kreider, eds. *Distributed Generation: The Power Paradigm for the New Millennium*. Washington, D.C.: CRC Press. 2001. p. 96.
41. Hoffmann. *Tomorrow's Energy*. pp. 53–55.
42. Romm, Joseph J. and Charles B. Curtis. "Mideast Oil Forever?" *The Atlantic Online*. April 1996.
43. "Shell Oil Chairman and President Steve Miller Addresses the Congressional Black Caucus." Shell Oil. June 29, 2001. *www.countonshell.com/news/relations/speeches/speech15.html*
44. "When Green Begets Green." *Business Week*. November 10, 1997. p. 98
45. Houghton. *Global Warming*. p. 203.
46. Ibid. p. 208; Hoffmann. *Tomorrow's Energy* p. 90; Seth Dunn. "Micropower: The Next Electrical Era." Worldwatch Paper 151. Washington, D.C.: Worldwatch Institute, July 2000.
47. Hoffmann. *Tomorrow's Energy*. p. 90;
48. Chambers. *Distributed Generation*. p. 124.
49. Ibid.
50. Ibid. p. 125.
51. Hoffmann. *Tomorrow's Energy*. p. 89.
52. Wagner, Andreas. German Wind Energy Association. "The Growth of Wind Energy in Europe—An Example of Successful Regulatory and Financial Incentives." Presentation to Windpower '99 Conference. Burlington, VT: American Wind Energy Association, June 21, 1999; Krohn, Søren et al., eds. "Windpower: DWTMA Annual Report. No. 25." March 2000. p. 3.
53. Wagner. "The Growth of Wind Energy in Europe." Larry Goldstein, John Mortensen, and David Trickett. *Grid-Connected Renewable-Electric Policies in the European Union*. Golden, CO: National Renewable Energy Laboratory (NREL), May 1999.
54. Report of the Renewable Energy Advisory Group. Energy Paper Number 60. UK Department of Trade and Industry, November 1992.
55. Lloyd and Hassan cited in Beursken, Jos. "Going to Sea: Wind Goes Offshore." Renewable Energy World, January–February 2000.

56. Hoffmann. *Tomorrow's Energy*. p. 89; Lal, M. "Measures for Reducing Climate Relevant Gas Emissions in India." Paper presented at an Indo-German seminar IIT. Delhi: October 29–31, 1996.
57. Silberman. "The Energy Web." p. 118.
58. Houghton. *Global Warming*. p. 203.
59. Hoffmann. *Tomorrow's Energy*. p. 88.
60. Dunn. "Hydrogen Futures." p. 32
61. Ibid. p. 214.
62. Ibid. pp. 96–97.
63. Ibid. p. 97.
64. Houghton. *Global Warming*. p. 205.
65. Romm and Curtis. "Mideast Oil Forever?"
66. Ibid.
67. Hoffmann. *Tomorrow's Energy*. p. 148.
68. Chambers. *Distributed Generation* p. 87; Hoffmann. *Tomorrow's Energy*. p. 155.
69. Hoffmann. *Tomorrow's Energy*. p. 6.
70. Dunn. "Hydrogen Futures." p. 32.
71. Fairley, Peter. "Power to the People." *Technology Review*. MIT Enterprise. May 2001.
72. "Distributed Generation: Understanding the Economics." An Arthur D. Little White Paper. 1999.
73. Chambers. *Distributed Generation*. p. 18.
74. Lloyd, Alan C. "The Power Plant in Your Basement." *Scientific American*. July 1999. p. 80.
75. Wald, Matthew L. "Energy to Count On." *The New York Times*. August 17, 1999. p. C1.; Romm Joseph J. *Cool Companies: How the Best Businesses Boost Profits and Productivity by Cutting Greenhouse Gas Emissions*. Washington, D.C.: Island Press, 1999.
76. Silberman. "The Energy Web." p. 117.
77. Chambers. *Distributed Generation*. p. 8.
78. Lloyd. "The Power Plant in Your Basement." p. 83.
79. Fairley. "Power to the People." p. 74.
80. Lloyd. "The Power Plant in Your Basement." p. 83.
81. Ibid.
82. Ibid.
83. Chambers. *Distributed Generation*. p. 13.
84. Chambers. *Distributed Generation*. pp. 5–6, 10.
85. IPCC. "Climate Change 2001: Impacts, Adaptation and Vulnerability." Summary for Policymakers. Working Group II Contribution to the Third Assessment Report of the IPCC. Geneva: February 19, 2001; U.N. Development Programme (UNDP), U.N. Department of Economic and Social Affairs (UNDESA), and World Energy Council (WEC). *World Assessment Report*. New York: 2000. pp. 116, 74–77, 86–90.
86. Fairley. "Power to the People." p. 74.
87. Chambers. *Distributed Generation*. p. 22.
88. Personal correspondence with Joel Smisher of the Rocky Mountain Institute. March 4, 2002.
89. Chambers. *Distributed Generation* p. 24.
90. Ibid.

91. "Distributed Generation: Understanding the Economics." An Arthur D. Little White Paper.
92. Chambers. *Distributed Generation.* p. 22.
93. Silberman. "The Energy Web." p. 120.
94. Borbely, Anne-Marie and Jan F. Kreider, eds. *Distributed Generation: The Power Paradigm for the New Millennium.* Washington, D.C.: CRC Press, 2001. p. 47.
95. Miller, Steven E. *Civilizing Cyberspace: Policy, Power, and the Information Superhighway.* New York: Addison-Wesley, 1996. pp. 44–45.
96. Fairley. "Power to the People." p. 77.
97. Silberman. "The Energy Web." p. 120.
98. Ibid.
99. Chambers. *Distributed Generation.* p. 25.
100. Yeager quoted in Silberman. "The Energy Web." p. 117.
101. Ibid. p. 121.
102. Ibid.
103. Ibid.
104. Ibid. p. 124.
105. Alderfer, R. Brent, M. Monika Eldridge and Thomas J. Starrs. "Making Connections: Case Studies of Interconnection Barriers and their Impact on Distributed Power Projects." NREL/SR-200-28053. Golden, CO: National Renewable Energy Laboratory, May 2000. p. iv.
106. Hawken, Lovins, and Lovins. Natural Capitalism. p. 26; Silberman. "The Energy Web." p. 120.
107. Silberman. "The Energy Web." p. 120; for further reading on current developments with hydrogen cars, see Jim Motavalli, *Forward Drive: The Race to Build "Clean" Cars for the Future.* San Francisco: Sierra Club Books, 2000.
108. Ibid.
109. "ZEVCO Unveils Fuel Cell Taxi, Shell UK Chief Says Company Is Into Hydrogen for Real." *Hydrogen & Fuel Cell Letter*, August 1998.
110. Mitchell, Marnie. "Balancing Movement with Management." *International Herald Tribune.* September 27, 2001.
111. Ibid.
112. U.S. Energy Information Administration. *Annual Energy Review*, 1997.
113. "Electricity Technology Roadmap: Powering Progress." 1999 Summary and Synthesis. Palo Alto, CA: EPRI, July 1999.
114. Mitchell. "Balancing Movement with Management."
115. Lovins, Amory B. and Brett D. Williams. "From Fuel Cells to a Hydrogen-based Economy." *Public Utilities Fortnightly.* Vol. 139, No. 4. February 15, 2001. p. 15.
116. Ibid. pp. 15–16.
117. Ibid. p. 16.
118. Ibid. p. 20.
119. Thomas, C. E. et al. "Fuel Options for the Fuel Cell Vehicle: Hydrogen, Methanol or Gasoline?" *International Journal of Hydrogen Energy.* Vol. 25. 2000. p. 552.
120. Ibid.
121. Ibid. p. 564.
122. Hoffmann. *Tomorrow's Energy.* p. 112.

123. Nahmias, Dave. "Fuel Choice for Fuel Cell Vehicles." Report to the Department of Energy from Its Hydrogen Technical Advisory Board. National Hydrogen Association, 1999.
124. Ibid.
125. Ibid.
126. Ibid.
127. Hoffmann. *Tomorrow's Energy*. p. 99.
128. Lovins and Williams. "From Fuel Cells to a Hydrogen-based Economy." p. 17.
129. Hoffmann. *Tomorrow's Energy*. p. 235.
130. Ibid. p. 242.
131. Swoboda, Frank. "Engines of Changes?" *The Washington Post*. January 8, 2002. p. E1. Truett, Richard. "GM Sticks to 2010 Target for Fuel Cell Vehicles." *Automotive News*. February 11, 2003.
132. Banerjee, Neela with Danny Hakim. "Administration Shifts Strategy on Auto Fuels." *The New York Times*. January 9, 2002. p. A1.
133. "Next Energy: Powering Michigan's Future." State of Michigan. *www.michigan.gov*
134. Ibid.
135. Meller, Paul. "Europe Pushes for Renewable Energy." *The New York Times*. October 16, 2002.
136. de Palacia, Loyola. "New and Original Ways for Hydrogen to Reach the Market." Speech to the High Level Group for Hydrogen and Fuel Cells. Brussels, October 10, 2002; "Commission Launches High Level Group on Hydrogen and Fuel Cells." The European Commission. Brussels, October 10, 2002.
137. Meller, Paul. "Europe Pushes for Renewable Energy." *The New York Times*. October 16, 2002.
138. "Fuel Cells and Hydrogen: The Path Forward." *The Fuel Cell Coalition*. September 5, 2002.
139. Bush, George W. "State of the Union." January 28, 2003.
140. Bush, George W. "State of the Union." January 28, 2003; "Staff Draft Summary: Title VIII—Hydrogen." US Senate Committee on Energy and Natural Resources. March 25, 2003.

REGLOBALIZATION FROM THE BOTTOM UP

1. Barlow, John Perry, "The Economy of Ideas: Rethinking Patents and Copyrights in the Digital Age." *Wired*. Vol. 2, No. 3, 1994. p. 22.
2. Thompson, Nicholas. "Reboot!" *Washington Monthly*. March 2000.
3. Chobham, Thomas. *Summa Confessorum*, ed. F. Broomfield. Paris: Louvain, 1968. p. 505, question XI, chapter 1.
4. Swarztrauber, Sayre A. *The Three-Mile Limit of Territorial Seas*. Annapolis, MD: Naval Institute Press, 1972. pp. 23–35.
5. Sorroos, Marvin S. "The International Commons: A Historical Perspective." *Environmental Review* 12, Spring 1988. p. 14.
6. Churchill, R. R. and A. V. Lowe. *The Law of the Sea*. Vol. 1. Oxford, England: Oxford University Press, 1983. p. 126.
7. Ibid.

8. Shapiro, Andrew L. *The Control Revolution: How the Internet Is Putting Individuals in Charge and Changing the World We Know*. New York: PublicAffairs, 1999. p. 210.
9. Rusk, David. *Inside Game, Outside Game*. Washington, D.C.: Brookings Institution, 1999. p. 25.
10. *Coming of Age: Trends and Achievements of Community-based Development Organizations.* Washington, D.C.: National Congress for Community Economic Development, 1999.
11. Ibid.
12. 1999 Annual Report. New Community Corporation. *www.newcommunity.org*
13. National Credit Union Association. *www.ncua.gov/news/cdcu/cdufact.html#cdcufact*
14. American Public Power Association. *www.appanet.org*
15. Ibid.
16. American Public Power Association. "Fact Sheet." April 1996.
17. Salpukas, Agis. "The Rebellion in 'Pole City.'" *The New York Times*. October 10, 1995. p. D1.
18. McKenzie, Evan. *Privatopia: Homeowner Associations and the Rise of Residential Private Government*. New Haven, CT: Yale University Press, 1996. p. 12.
19. Winokur, James L. "Choice, Consent, and Citizenship in Common Interest Communities." In Stephen E. Barton and Carol J. Silverman, eds., *Common Interest Communities: Private Governments and the Public Interest*. Berkeley: Institute of Governmental Studies, 1994. p. 88.
20. Nelson cited in McKenzie. *Privatopia*. pp. 176–177.
21. Draft from the National Center for Economic and Security Alternatives. 2001. *www.ncba.org*
22. International Co-operative Alliance website. 2001. *www.coop.org/ica/ica-intro.html*
23. National Cooperative Bank. *NCB Co-op 100*. 2001. *www.ncb.com*
24. Hazen, Paul. "The Next Cooperative Wave." Cooperative Business Journal. National Cooperative Business Association. 2001. *www.cooperative.org/prescols.cfm?colid=15*
25. Williamson, Thad, David Imbroscio and Gar Alperovitz. "Making a Place for Community: A Policy Primer for the New Century." To be published by Routledge Press. p. 379
26. Howard, Ted and Kristin Rusch. "Community Asset Building and Sustainable Resource Management: A Preliminary Survey." (Unpublished report.) College Park, MD: National Center for Economic Security Alternatives. February 2001. p. 6. For more information on community-based nonprofit economic models, contact The Democracy Collaborative at the University of Maryland: *info@DemocracyCollaborative.org*
27. Ibid. p. 13.
28. Ibid. p. 21.
29. Ibid. p. 39.
30. Touchstone Energy Cooperative. *www.touchstoneenergy.com*
31. Energy Cooperatives Network. *www.energy-co-op.net*
32. Ibid.
33. Ibid.
34. Ibid.
35. "Chicago Area Homeowners to Test Residential Fuel Cells." Electric Power Research Institute (EPRI). May 10, 2001.

36. "It's Time to Get Smarter about Energy." Community Energy Cooperative. *www.energy cooperative.net*
37. Energy Cooperatives Network. *www.energy-co-op.net*
38. Miller, Steven E. *Civilizing Cyberspace: Policy, Power, and the Information Superhighway*. New York: Addison-Wesley, 1996. p. 206.
39. "Total Midyear Population for the World: 1950–2050." International Data Base. U.S. Bureau of the Cersus. May 10, 2000.
40. Merchant, Khozen. "World Heads for Grotesque Inequalities." *Financial Times*. July 16, 1996. p. 4; United Nations Development Program (UNDP). *Human Development Report*, 1998. New York: Oxford University Press, 1998.
41. Taylor, Robert. "Market Fallout Will Lift Jobless Total: World Unemployment—Third of All Workers Affected Says ILO Report." *Financial Times*. September 24, 1998. p. 8.
42. Crosette, Barbara. "Hope and Pragmatism for U.N. Cities Conference." *The New York Times*. June 3, 1996. p. A3.
43. United Nations Development Program (UNDP). *Human Development Report*, 1998.
44. UNDP. *Human Development Report*, 1996. New York: Oxford University Press, 1996.
45. "Electricity Technology Roadmap: Powering Progress." 1999 Summary and Synthesis. Palo Alto, CA: EPRI, July 1999. pp. 96–97.
46. Starr, Chauncey. "Sustaining the Human Environment: The Next Two Hundred Years." In Jesse H. Ausubel and H. Dalle Langford, eds. *Technological Trajectories and the Human Environment*. Washington, D.C.: National Academy Press, 1997. p. 192.
47. "Electricity Technology Roadmap: Powering Progress." p. 98.
48. Ibid.
49. Ibid. p. 68.
50. Ibid.
51. Ibid.
52. Ibid.
53. Ibid.
54. Ibid. p. 103.
55. Ibid. p. 104.
56. Ibid. p. 103.
57. "Voodoo Economics: A Thumbnail Sketch of the Global Finance System and Just Where the World Bank Fits in It." *New Internationalist*, Issue 214, December 1990.
58. Ibid.
59. Easley, Dale H. "The OPEC Oil Embargo and Third-World Debt." July 17, 2000. *www.uno.edu/~gege/Easley/Essays*.
60. "High Prices Hurt Poor Countries More Than Rich." Paris: March 20, 2000. *www.iea.org/new/releases/2000/oilprice.html*
61. Ibid.
62. Ibid.
63. Annan, Kofi A. "Where the High Oil Price Really Hurts." *International Herald Tribune*, October 3, 2000.
64. Roodman, David Malin. "The Third World Debt Crisis: Facts and Myths." *Worldwatch Paper 155: Still Waiting for the Jubilee: Pragmatic Solutions for the Debt Crisis*. April 26, 2001.
65. Ibid.

66. "Forgive and Forget" Won't Fix Third World Debt." April 26, 2001. *www.worldwatch.org/alerts/010426.html*
67. Roodman, "The Third World Debt Crisis."
68. Ratzel cited in Kern, Stephen. *The Culture of Time and Space*, 1880–1918. Cambridge, MA: Harvard University Press, 1983. p. 224.
69. Ratzel, Friedrich. *Das Meer als Quelle der Volergrosse*. Munich: 1900. pp. 1, 5; Kern. *The Culture of Time and Space*, 1880–1918. p. 226.
70. Mackinder, Sir Halford. *Democratic Ideas and Reality*. New York: W. W. Norton, 1962.
71. Spykman, Nicholas J. *The Geography of Peace*. New York: Harcourt, Brace, 1944. p. 5.
72. Vernadsky quoted in Serafin, Rafal. "Noosphere, Gaia, and the Science of the Biosphere." *Environmental Ethics* 10, Summer 1988. p. 124.
73. Lovelock, James, in William Irvin Thompson, ed., *Gaia: A New Way of Knowing*. New York: Lindisfarne Press, 1988. pp. 87–88.
74. Lovelock, James. *Gaia: A New Look at Life on Earth*. Oxford, England: Oxford University Press, 1979. p. 72.
75. Ibid.
76. Ibid. p. 73.
77. Ibid. p. 74.
78. Lovelock. *The Ages of Gaia*. New York: W. W. Norton, 1988. pp. 33–34.

BIBLIOGRAPHY

Adams, Richard Newbold. *Energy and Structure: A Theory of Social Power*. Austin: University of Texas Press, 1975.

Ahmad, Q. K. et al. "Summary for Policy Makers: Climate Change 2001: Impacts, Adaptation, and Vulnerability." A Report of Working Group II of the Intergovernmental Panel on Climate Change. IPCC. February 2001.

Albritton, Daniel L. et al. "Summary for Policy Makers: Climate Change 2001: The Scientific Basis." A Report of Working Group I of the Intergovernmental Panel on Climate Change. IPCC. February 2001.

Anderson, Robert O. *Fundamentals of the Petroleum Industry*. Norman: University of Oklahoma Press, 1984.

Armstrong, Karen. *Islam: A Short History*. New York: Modern Library, 2000.

Banuri, Tariq et al. "Summary for Policy Makers: Climate Change 2001: Mitigation." A Report of Working Group III of the Intergovernmental Panel on Climate Change. IPCC. February–March 2001.

Barber, Benjamin R. *Jihad Versus McWorld*. New York: Ballantine Books, 2001.

Bateson, Gregory. *Steps to an Ecology of Mind: Collected Essays in Anthropology, Psychiatry, Evolution, and Epistemology*. Northvale, NJ: Jason Arson Inc., 1987.

Beniger, James R. *The Control Revolution*. Cambridge, MA: Harvard University Press, 1986.

Blum, Harold F. *Time's Arrow and Evolution*. Princeton, NJ: Princeton University Press, 1968.

Borbely, Anne-Marie, and Jan F. Kreider, eds. *Distributed Generation: The Power Paradigm for the New Millennium*. Washington, D.C.: CRC Press, 2001.

Catton, William R., Jr. *Overshoot: The Ecological Basis of Revolutionary Change*. Urbana: University of Illinois Press, 1980.

Chambers, Ann. *Distributed Generation: A Nontechnical Guide*. Tulsa, OK: PennWell, 2001.

Chandler, Alfred D., Jr. *The Visible Hand: The Managerial Revolution in American Business*. Cambridge, MA: Belknap Press of Harvard University Press, 1977.

Chomsky, Noam. *Rogue States: The Rule of Force in World Affairs*. Cambridge, MA: South End Press, 2000.

Clark, Wilson. *Energy for Survival*. Garden City, NY: Doubleday/Anchor Books, 1975.

Cochrane, Willard. *Development of American Agriculture: A Historical Analysis*, 2nd ed. Minneapolis: University of Minnesota Press, 1993.

Condorcet, Marie, Marquis de. *Outlines of an Historical View of the Progress of the Human Mind*. London: J. Johnson, 1795.

Corn, Joseph J., ed. *Imagining Tomorrow: History, Technology, and the American Future*. Cambridge, MA: MIT Press, 1986.

Davidson, Lawrence. *Islamic Fundamentalism*. Westport, CT: Greenwood Press, 1998.

Deffeyes, Kenneth S. *Hubbert's Peak: The Impending World Oil Shortage*. Princeton, NJ: Princeton University Press, 2001.

Dunn, Seth. "Hydrogen Futures: Toward a Sustainable Energy System," *Worldwatch Paper 157*. Washington, DC: Worldwatch Institute, August 2001.

———. "Micropower: The Next Electrical Era," *Worldwatch Paper* 151. Washington, D.C.: Worldwatch Institute, July 2000.

Economides, Michael, and Ronald Oligney. *The Color of Oil: The History, the Money and the Politics of the World's Biggest Business*. Katy, TX: Round Oak Publishing, 2000.

Falkenrath, Richard A., Robert D. Newman, and Bradley A. Thayer. *America's Achilles' Heel: Nuclear, Biological, and Chemical Terrorism and Covert Attack*. Cambridge, MA: MIT Press, 1998.

Farb, Peter. *Humankind*. Boston: Houghton Mifflin, 1978.

Fergosi, Paul. *Jihad in the West: Muslim Conquests from the 7th to the 21st Centuries*. Amherst, NY: Prometheus Books, 1998.

Frank, Tenney. *An Economic Survey of Ancient Rome*, Volume V: *Rome and Italy of the Empire*. Baltimore: Johns Hopkins Press, 1940.

Gause, F. Gregory, III. *Oil Monarchies: Domestic and Security Challenges in the Arab Gulf States*. New York: Council on Foreign Relations Press, 1994.

Gever, John, David Skole, and Robert K. Kaufmann. *Beyond Oil: The Threat to Food and Fuel in the Coming Decades*. Niwot: University Press of Colorado, 1991.

Gibbon, Edward. *The Decline and Fall of the Roman Empire*. New York: Modern Library, 1776–1788.

Goodman, David et al. *From Farming to Biotechnology: A Theory of Agro-Industrial Development*. New York: Basil Blackwell, 1987.

Hawken, Paul, Amory Lovins, and L. Hunter Lovins. *Natural Capitalism: Creating the Next Industrial Revolution*. Boston: Little, Brown and Company, 1999.

Hirsh, Richard F. *Power Loss: The Origins of Deregulation and Restructuring in the American Electric Utility System*. Cambridge, MA: MIT Press, 1999.

Hoffmann, Peter. *Tomorrow's Energy: Hydrogen, Fuel Cells, and the Prospect for a Cleaner Planet*. Cambridge, MA: MIT Press, 2001.

Houghton, John. *Global Warming: The Complete Briefing*. Cambridge, MA: Cambridge University Press, 1997.

Hoveyda, Fereydoun. *The Broken Crescent: The Threat of Militant Islamic Fundamentalism*. Westport, CT: Praeger, 1998.

Huband, Mark. *Warriors of the Prophet: The Struggle for Islam*. Boulder, CO: Westview Press, 1998.

Hughes, Donald J. *Ecology in Ancient Civilizations*. Albuquerque: University of New Mexico Press, 1975.

Hughes, Thomas P. *Networks of Power: Electrication in Western Society, 1880–1930*. Baltimore: Johns Hopkins University Press, 1983.

Huntington, Samuel P. *The Clash of Civilizations and the Remaking of World Order*. New York: Simon and Schuster, 1996.

Johnson, Chalmers. *Blowback: The Costs and Consequences of American Empire*. New York: Metropolitan Books, 2000.

Jones, A. H. M. *The Roman Economy: Studies in Ancient Economic and Administrative History*. Oxford: Basil Blackwell, 1974.

Juergensmeyer, Mark. *Terror in the Mind of God: The Global Rise of Religious Violence*. Berkeley: University of California Press, 2000.

Khavari, Farid A. *Oil and Islam: The Ticking Bomb*. Malibu, CA: Roundtable Publishers, 1990.

Kranzberg, Melvin, and Carroll W. Pursell Jr., eds. *Technology in Western Civilization: The Emergence of Modern Industrial Society: Earliest Times to 1900*. Vols. I, II. New York: Oxford University Press, 1967.

Landes, David S. *Revolution in Time*. Cambridge, MA: Harvard University Press, 1983.

Levy, Jean-Philippe. *The Economic Life of the Ancient World*. Translated by John G. Biram. Chicago: University of Chicago Press, 1967.

Lovins, Amory B., and L. Hunter Lovins. *Brittle Power: Energy Strategy for National Security*. Andover, MA: Brick House Publishing Co., Inc., 2002.

Marcuse, Herbert. *One-Dimensional Man: Studies in the Ideology of Advanced Industrial Society*. Boston: Beacon Press, 1964.

McNeill, William H. *Plagues and Peoples*. Garden City, NY: Anchor/Doubleday, 1976.

Melko, Matthew. *The Nature of Civilizations*. Boston: Porter Sargent, 1969.

Miller, G. Tyler, Jr. *Energetics, Kinetics and Life*. Belmont, CA: Wadsworth, 1971.

Miller, Steven E. *Civilizing Cyberspace: Policy, Power, and the Information Superhighway*. New York: Addison-Wesley, 1996.

Moaddel, Mansoor, and Kamran Talattof, eds. *Contemporary Debates in Islam: An Anthology of Modernist and Fundamentalist Thought*. New York: St. Martin's Press, 2000.

Motavalli, Jim. *Forward Drive: The Race to Build "Clean" Cars for the Future*. San Francisco: Sierra Club Books, 2000.

Mowbray, A. Q. *Road to Ruin*. Philadelphia: Lippincott, 1969.

Mumford, Lewis. *Technics and Civilization*. New York: Harcourt, Brace, 1934.

Noble, David F. *America by Design: Science, Technology, and the Rise of Corporate Capitalism*. New York: Alfred A. Knopf, 1977.

———. *Forces of Production*. New York: Oxford University Press, 1984.

Odum, Howard T. *Environment, Power, and Society*. Chapel Hill: University of North Carolina Press, 1971.

Philander, S. George. *Is the Temperature Rising? The Uncertain Science of Global Warming*. Princeton, NJ: Princeton University Press, 1998.

Pipes, Daniel. *In the Path of God: Islam and Political Power*. New York: Basic Books, 1983.

Ponting, Clive. *A Green History of the World: The Environment and the Collapse of Great Civilizations*. New York: Penguin Books, 1991.

Quigley, Carroll. *The Evolution of Civilizations*. Indianapolis: Liberty Press, 1961.

Rashid, Ahmed. *Taliban: Militant Islam, Oil and Fundamentalism in Central Asia*. New Haven, CT: Yale University Press, 2000.

Redman, Charles L. *Human Impact on Ancient Environments*. Tucson: University of Arizona Press, 1999.

Rifkin, Jeremy. *The Age of Access: The New Culture of Hypercapitalism Where All of Life Is a Paid-For Experience*. New York: Tarcher/Putnam, 2000.

———. *Biosphere Politics: A New Consciousness for a New Century*. New York: Crown Publishers, 1991.

———. *The Biotech Century: Harnessing the Gene and Remaking the World*. New York: Tarcher/Putnam, 1998.

———. *The End of Work: The Decline of the Global Labor Force and the Dawn of the Post-Market Era*. New York: Tarcher/Putnam, 1995.

———. *Entropy Into the Greenhouse World*. New York: Bantam Books, 1989.

———. *Time Wars: The Primary Conflict in Human History*. New York: Henry Holt and Company, 1987.

Romm, Joseph J. *Cool Companies: How the Best Businesses Boost Profits and Productivity by Cutting Greenhouse Gas Emissions*. Washington, D.C.: Island Press, 1999.

Rusk, David. *Inside Game, Outside Game*. Washington, D.C.: Brookings Institution, 1999.

Shadid, Anthony. *Legacy of the Prophet: Despots, Democrats, and the New Politics of Islam*. Boulder, CO: Westview Press, 2001.

Schneider, K. R. Autokind vs. Mankind. New York: Schoken, 1972.

Schrödinger, Erwin. *What Is Life?* New York: Macmillan, 1947.

Shapiro, Andrew L. *The Control Revolution: How the Internet Is Putting Individuals in Charge and Changing the World We Know*. New York: Public Affairs, 1999.

Soddy, Frederick. *Matter and Energy*. Home University Series. London: Oxford University, 1912.

Soto, Hernando de. *The Mystery of Capital*. New York: Basic Books, 2000.

Spengler, Oswald. *The Decline of the West*. Translated by Charles Francis Atkinson. New York: Modern Library, 1962.

Stern, Jessica. *The Ultimate Terrorists*. Cambridge, MA: Harvard University Press, 1999.

Swarztrauber, Sayre A. *The Three-Mile Limit of Territorial Seas*. Annapolis, MD: Naval Press Institute Press, 1972.

Tainter, Joseph A. *The Collapse of Complex Societies*. Cambridge, MA: Cambridge University Press, 1988.

Taylor, Frederick. *The Principles of Scientific Management*. New York: W. W. Norton, 1947.

Thomas, William L., ed. *Man's Role in Changing the Face of the Earth*. Chicago: Wenner-Gren Foundation for Anthropological Research and the National Science Foundation by the University of Chicago Press, 1956.

Tibi, Bassam. *The Challenge of Fundamentalism*. Berkeley: University of California Press, 1998.

Tichi, Cecelia. *Shifting Gears*. Chapel Hill: University of North Carolina Press, 1987.

Verne, Jules. *The Mysterious Island*. Translated by W. H. G. Kingston. New York: The Limited Editions Club at the Garamond Press, 1959.

Victor, David G. *The Collapse of the Kyoto Protocol and the Struggle to Slow Global Warming*. Princeton, NJ: Princeton University Press, 2001.

White, Leslie A. *The Evolution of Culture*. New York: McGraw-Hill Company, 1959.

———. *The Science of Culture: A Study of Man and Civilization*. New York: Farrar, Straus, and Company, 1949.

Whitehead, Alfred North. *Science and the Modern World*. New York: New American Library, 1925.

Wilkinson, Richard G. *Poverty and Progress: An Ecological Perspective on Economic Development*. New York: Praeger Publishers, 1973.

Winkless, Nels, III, and Iben Browning. *Climate and the Affairs of Men*. New York: Harper's Magazine Press, 1975.

Yergin, Daniel. *The Prize: The Epic Quest for Oil, Money and Power*. New York: Simon and Schuster, 1992.

Youngquist, Walter. *GeoDestinies: The Inevitable Control of Earth Resources Over Nations and Individuals*. Portland, OR: National Book Company. 1997.

INDEX

ABOUT THE AUTHOR

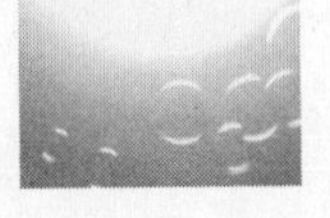

Jeremy Rifkin is an internationally renowned social critic and the bestselling author of *The End of Work, The Biotech Century*, and *The Age of Access*, each of which has been translated into more than fifteen languages. He is president of the Foundation on Economic Trends in Washington, D.C. Since 1994, Mr. Rifkin has been a fellow at the Wharton School's Executive Education Program, where he lectures to CEOs and senior corporate management from around the world on new trends in science and technology and their impacts on the global economy, society, and the environment. He is also an advisor to heads of state and government officials in a number of countries. Mr. Rifkin's monthly column on global issues appears in many of the world's leading newspapers and magazines. He lives with his wife, Carol Grunewald, in Washington, D.C.